ELECTRONICS PROJECTS
READY-REFERENCE

BOOKS by JOHN MARKUS

ELECTRONIC CIRCUITS MANUAL

ELECTRONICS DICTIONARY

ELECTRONICS FOR COMMUNICATION ENGINEERS

ELECTRONICS FOR ENGINEERS

ELECTRONICS MANUAL FOR RADIO ENGINEERS

ELECTRONICS AND NUCLEONICS DICTIONARY

GUIDEBOOK OF ELECTRONIC CIRCUITS

HANDBOOK OF ELECTRONIC CONTROL CIRCUITS

HANDBOOK OF INDUSTRIAL ELECTRONIC CIRCUITS

HANDBOOK OF INDUSTRIAL ELECTRONIC CONTROL CIRCUITS

HOW TO GET AHEAD IN THE TELEVISION AND
RADIO SERVICING BUSINESS

MODERN ELECTRONIC CIRCUITS REFERENCE MANUAL

SOURCEBOOK OF ELECTRONIC CIRCUITS

TELEVISION AND RADIO REPAIRING

WHAT ELECTRONICS DOES

ELECTRONICS PROJECTS READY-REFERENCE

JOHN MARKUS

Consultant, McGraw-Hill Book Company
Senior Member, Institute of Electrical and Electronics Engineers

McGRAW-HILL BOOK COMPANY

New York St. Louis San Francisco Auckland Bogotá Hamburg
Johannesburg London Madrid Mexico Montreal New Delhi
Panama Paris São Paulo Singapore Sydney Tokyo Toronto

Library of Congress Cataloging in Publication Data

Markus, John, date.
   Electronics projects ready-reference.

   Includes index.
   1. Electronics—Amateurs' manuals.   2. Electronic
circuits—Amateurs' manuals.   I. Title.
TK9965.M28   621.3815'3   82-108
ISBN 0-07-040459-3   AACR2

Portions of this book originally appeared in *Modern Electronic
Circuits Reference Manual*, copyright © 1980 by McGraw-Hill, Inc.

1234567890   HD HD   898765432

ISBN 0-07-040459-3

Contents

Preface

Electronics Projects Ready-Reference is one of five books in the Ready-Reference series. These books are the product of cover-to-cover searching of back issues of U.S. and foreign electronics periodicals, the published literature of electronics manufacturers, and recent electronics books, together filling well over 100 feet of shelving. This same search would take weeks or even months at a large engineering library, plus the time required to write for manufacturer literature and locate elusive sources.

Each circuit has type numbers or values of all significant components, an identifying title, a concise description, performance data, and suggested applications. At the end of each description is a citation giving the title of the original article or book, its author, and the exact location of the circuit in the original source.

To find a desired circuit quickly, start with the alphabetically arranged table of contents at the front of the book. Note the chapters most likely to contain the desired type of circuit, and look in these first. If a quick scan does not locate the exact circuit desired, use the index at the back of the book. Here the circuits are indexed in depth under the different names by which they may be known. Cross-references in the index aid searching. The author index will often help find related circuits after one potentially useful circuit is found, because authors tend to specialize in certain circuits.

To the original publications cited and their engineering authors and editors should go major credit for making this book possible. The diagrams have been reproduced directly from the original source articles, by permission of the publisher in each case.

Abbreviations Used

A	ampere	CRO	cathode-ray oscilloscope
AC	alternating current		
AC/DC	AC or DC	CROM	control and read-only memory
A/D	analog-to-digital		
ADC	analog-to-digital converter	CRT	cathode-ray tube
		CT	center tap
A/D, D/A	analog-to-digital, or digital-to-analog	CW	continuous wave
		D/A	digital-to-analog
ADP	automatic data processing	DAC	digital-to-analog converter
AF	audio frequency	dB	decibel
AFC	automatic frequency control	dBC	C-scale sound level in decibels
AFSK	audio frequency-shift keying	dBm	decibels above 1 mW
		dBV	decibels above 1 V
AFT	automatic fine tuning	DC	direct current
		DC/DC	DC to DC
AGC	automatic gain control	DCTL	direct-coupled transistor logic
Ah	ampere-hour	diac	diode AC switch
ALU	arithmetic-logic unit	DIP	dual in-line package
AM	amplitude modulation	DMA	direct memory access
AM/FM	AM or FM	DMM	digital multimeter
AND	type of logic circuit	DPDT	double-pole double-throw
AVC	automatic volume control	DPM	digital panel meter
b	bit	DPST	double-pole single-throw
BCD	binary-coded decimal		
BFO	beat-frequency oscillator	DSB	double sideband
		DTL	diode-transistor logic
b/s	bit per second	DTL/TTL	DTL or TTL
C	capacitance; capacitor	DUT	device under test
		DVM	digital voltmeter
°C	degree Celsius; degree Centigrade	DX	distance reception; distant
CATV	cable television	EAROM	electrically alterable ROM
CB	citizens band		
CCD	charge-coupled device	EBCDIC	extended binary-coded decimal interchange code
CCTV	closed-circuit television	ECG	electrocardiograph
cm	centimeter	ECL	emitter-coupled logic
CML	current-mode logic	EDP	electronic data processing
CMOS	complementary MOS		
CMR	common-mode rejection	EKG	electrocardiograph
		EMF	electromotive force
CMRR	common-mode rejection ratio	EMI	electromagnetic interference
cm²	square centimeter	EPROM	erasable PROM
coax	coaxial cable	ERP	effective radiated power
COHO	coherent oscillator		
COR	carrier-operated relay	ETV	educational television
COS/MOS	complementary-symmetry MOS (same as CMOS)	eV	electronvolt
		EVR	electronic video recording
CPU	central processing unit	EXCLUSIVE-OR	type of logic circuit
		EXCLUSIVE-NOR	type of logic circuit
CR	cathode ray		

F	farad		
°F	degree Fahrenheit		
FET	field-effect transistor		
FIFO	first-in first-out		
FM	frequency modulation		
4PDT	four-pole double-throw		
4PST	four-pole single-throw		
FS	full scale		
FSK	frequency-shift keying		
ft	foot		
ft/min	foot per minute		
ft/s	foot per second		
ft²	square foot		
F/V	frequency-to-voltage		
F/V, V/F	frequency-to-voltage, or voltage-to-frequency		
G	giga- (10^9)		
GHz	gigahertz		
G-M tube	Geiger-Mueller tube		
h	hour		
H	henry		
HF	high frequency		
HFO	high-frequency oscillator		
hp	horsepower		
Hz	hertz		
IC	integrated circuit		
IF	intermediate frequency		
IGFET	insulated-gate FET		
IMD	intermodulation distortion		
IMPATT	impact avalanche transit time		
in	inch		
in/s	inch per second		
in²	square inch		
I/O	input/output		
IR	infrared		
JFET	junction FET		
k	kilo- (10^3)		
K	kilohm (,000 ohms); kelvin		
kA	kiloampere		
kb	kilobit		
keV	kiloelectronvolt		
kH	kilohenry		
kHz	kilohertz		
km	kilometer		
kV	kilovolt		
kVA	kilovoltampere		
kW	kilowatt		
kWh	kilowatthour		
L	inductance; inductor		
LASCR	light-activated SCR		

LASCS	light-activated SCS
LC	inductance-capacitance
LCD	liquid crystal display
LDR	light-dependent resistor
LED	light-emitting diode
LF	low frequency
LIFO	last-in first-out
lm	lumen
LO	local oscillator
logamp	logarithmic amplifier
LP	long play
LSB	least significant bit
LSI	large-scale integration
m	meter; milli- (10^{-3})
M	mega- (10^6); meter (instrument); motor
mA	milliampere
Mb	megabit
MF	medium frequency
mH	millihenry
MHD	magnetohydro-dynamics
MHz	megahertz
mi	mile
mike	microphone
min	minute
mm	millimeter
modem	modulator-demodulator
mono	monostable
MOS	metal-oxide semiconductor
MOSFET	metal-oxide semiconductor FET
MOST	metal-oxide semiconductor transistor
MPU	microprocessing unit
ms	millisecond
MSB	most significant bit
MSI	medium-scale integration
m^2	square meter
μ	micro- (10^{-6})
μA	microampere
μF	microfarad
μH	microhenry
μm	micrometer
μP	microprocessor
μs	microsecond
μV	microvolt
μW	microwatt
mV	millivolt
MVBR	multivibrator
mW	milliwatt
n	nano- (10^{-9})
N	negative
nA	nanoampere
NAB	National Association of Broadcasters
NAND	type of logic circuit
nF	nanofarad
nH	nanohenry

NMOS	N-channel MOS
NOR	type of logic circuit
NPN	negative-positive-negative
NPNP	negative-positive-negative-positive
NRZ	nonreturn-to-zero
NRZI	nonreturn-to-zero-inverted
ns	nanosecond
NTSC	National Television System Committee
nV	nanovolt
nW	nanowatt
OEM	original equipment manufacturer
opamp	operational amplifier
OR	type of logic circuit
p	pico- (10^{-12})
P	peak; positive
pA	picoampere
PA	public address
PAL	phase-alternation line
PAM	pulse-amplitude modulation
PC	printed circuit
PCM	pulse-code modulation
PDM	pulse-duration modulation
PEP	peak envelope power
pF	picofarad
PF	power factor
phono	phonograph
PIN	positive-intrinsic-negative
PIV	peak inverse voltage
PLL	phase-locked loop
PM	permanent magnet; phase modulation
PMOS	P-channel MOS
PN	positive-negative
PNP	positive-negative-positive
PNPN	positive-negative-positive-negative
pot	potentiometer
P-P	peak-to-peak
PPI	plan-position indicator
PPM	parts per million; pulse-position modulation
preamp	preamplifier
PRF	pulse repetition frequency
PROM	programmable ROM
PRR	pulse repetition rate
ps	picosecond
PSK	phase-shift keying
PTT	push to talk
PUT	programmable UJT
pW	picowatt
PWM	pulse-width modulation
Q	quality factor

QRP	low-power amateur radio
R	resistance; resistor
RAM	random-access memory
RC	resistance-capacitance
RF	radio frequency
RFI	radio-frequency interference
RGB	red/green/blue
RIAA	Recording Industry Association of America
RLC	resistance-inductance-capacitance
RMS	root-mean-square
ROM	read-only memory
rpm	revolution per minute
RTL	resistor-transistor logic
RTTY	radioteletype
RZ	return-to-zero
s	second
SAR	successive-approximation register
SAW	surface acoustic wave
SCA	Subsidiary Communications Authorization
scope	oscilloscope
SCR	silicon controlled rectifier
SCS	silicon controlled switch
S-meter	signal-strength meter
S/N	signal-to-noise
SNR	signal-to-noise ratio
SPDT	single-pole double-throw
SPST	single-pole single-throw
SSB	single sideband
SSI	small-scale integration
SSTV	slow-scan television
SW	shortwave
SWL	shortwave listener
SWR	standing-wave ratio
sync	synchronizing
T	tera- (10^{12})
TC	temperature coefficient
THD	total harmonic distortion
TR	transmit-receive
TRF	tuned radio frequency
triac	triode AC semiconductor switch
TTL	transistor-transistor logic

TTY	teletypewriter	V	volt	VSWR	voltage standing-wave ratio
TV	television	VA	voltampere	VTR	videotape recording
TVI	television interference	VAC	volts AC	VTVM	vacuum-tube voltmeter
TVT	television typewriter	VCO	voltage-controlled oscillator	VU	volume unit
TWX	teletypewriter exchange service	VDC	volts DC	VVC	voltage-variable capacitor
UART	universal asynchronous receiver-transmitter	V/F	voltage-to-frequency	VXO	variable-frequency crystal oscillator
		VFO	variable-frequency oscillator	W	watt
UHF	ultrahigh frequency	VHF	very high frequency	Wh	watthour
UJT	unijunction transistor	VLF	very low frequency	WPM	words per minute
		VMOS	vertical metal-oxide semiconductor	WRMS	watts RMS
UPC	universal product code	VOM	volt-ohm-milliammeter	Ws	wattsecond
UPS	uninterruptible power system	VOX	voice-operated transmission	Z	impedance
		VRMS	volts RMS		

Abbreviations on Diagrams. Some foreign publications, including *Wireless World,* shorten the abbreviations for units of measure on diagrams. Thus, μ after a capacitor value represents μF, n is nF, and p is pF. With resistor values, k is thousand ohms, M is megohms, and absence of a unit of measure is ohms. For a decimal value, the letter for the unit of measure is sometimes placed at the location of the decimal point. Thus, 3k3 is 3.3 kilohms or 3,300 ohms, 2M2 is 2.2 megohms, 4μ7 is 4.7 μF, 0μ1 is 0.1 μF, and 4n7 is 4.7 nF.

Semiconductor Symbols Used

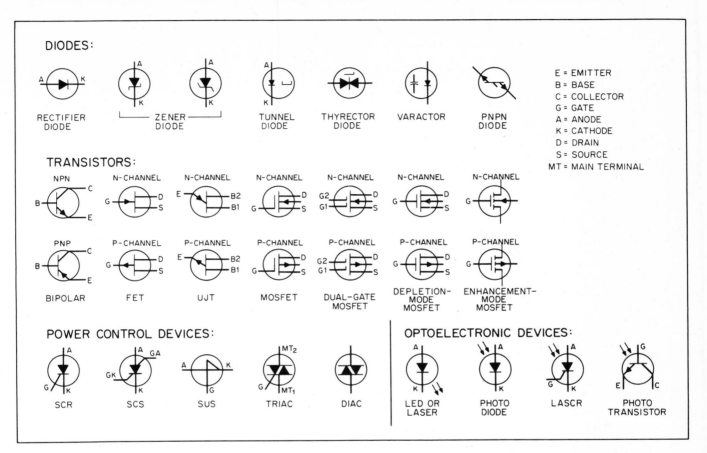

DIODES:

RECTIFIER DIODE · ZENER DIODE · TUNNEL DIODE · THYRECTOR DIODE · VARACTOR · PNPN DIODE

E = EMITTER
B = BASE
C = COLLECTOR
G = GATE
A = ANODE
K = CATHODE
D = DRAIN
S = SOURCE
MT = MAIN TERMINAL

TRANSISTORS:

NPN · N-CHANNEL · N-CHANNEL · N-CHANNEL · N-CHANNEL · N-CHANNEL · N-CHANNEL

PNP · P-CHANNEL · P-CHANNEL · P-CHANNEL · P-CHANNEL · P-CHANNEL · P-CHANNEL

BIPOLAR · FET · UJT · MOSFET · DUAL-GATE MOSFET · DEPLETION-MODE MOSFET · ENHANCEMENT-MODE MOSFET

POWER CONTROL DEVICES:

SCR · SCS · SUS · TRIAC · DIAC

OPTOELECTRONIC DEVICES:

LED OR LASER · PHOTO DIODE · LASCR · PHOTO TRANSISTOR

The commonest forms of the basic semiconductor symbols are shown here. Leads are identified where appropriate, for convenient reference. Minor variations in symbols, particularly those from foreign sources, can be recognized by comparing with these symbols while noting positions and directions of solid arrows with respect to other symbol elements.

Omission of the circle around a symbol has no significance. Arrows are sometimes drawn open instead of solid. Thicker lines and open rectangles in some symbols on diagrams have no significance. Orientation of symbols is unimportant; artists choose the position that is most convenient for making connections to other parts of the circuit. Arrow lines outside optoelectronic symbols indicate the direction of light rays.

On some European diagrams, the position of the letter k gives the location of the decimal point for a resistor value in kilohms. Thus, 2k2 is 2.2K or 2,200 ohms. Similarly, a resistance of 1R5 is 1.5 ohms, 1M2 is 1.2 megohms, and 3n3 is 3.3 nanofarads.

Substitutions can often be made for semiconductor and IC types specified on diagrams. Newer components, not available when the original source article was published, may actually improve the performance of a particular circuit. Electrical char-

acteristics, terminal connections, and such critical ratings as voltage, current, frequency, and duty cycle, must of course be taken into account if experimenting without referring to substitution guides.

Semiconductor, integrated-circuit, and tube substitution guides can usually be purchased at electronic parts supply stores.

Not all circuits give power connections and pin locations for ICs, but this information can be obtained from manufacturer data sheets. Alternatively, browsing through other circuits may turn up another circuit on which the desired connections are shown for the same IC.

When looking down at the top of an actual IC, numbering normally starts with 1 for the first pin *counterclockwise* from the notched or otherwise marked end and continues sequentially. The highest number is therefore next to the notch on the other side of the IC, as illustrated in the sketches below. (*Actual positions* of pins are rarely shown on schematic diagrams.)

Addresses of Sources Used

In the citation at the end of each abstract, the title of a magazine is set in italics. The title of a book or report is placed in quotes. Each source title is followed by the name of the publisher of the original material, plus city and state. Complete mailing addresses of all sources are given below, for the convenience of readers who want to write to the original publisher of a particular circuit. When writing, give the complete citation, exactly as in the abstract.

Books can be ordered from their publishers, after first writing for prices of the books desired. Some electronics manufacturers also publish books and large reports for which charges are made. Many of the books cited as sources in this volume are also sold by bookstores and by electronics supply firms. Locations of these firms can be found in the YELLOW PAGES of telephone directories under headings such as "Electronic Equipment and Supplies" or "Television and Radio Supplies and Parts."

Only a few magazines have back issues on hand for sale, but most magazines will make copies of a specific article at a fixed charge per page or per article. When you write to a magazine publisher for prices of back issues or copies, give the *complete* citation, *exactly* as in the abstract. Include a stamped self-addressed envelope to make a reply more convenient.

If certain magazines consistently publish the types of circuits in which you are interested, use the addresses below to write for subscription rates.

American Microsystems, Inc., 3800 Homestead Rd., Santa Clara, CA 95051

Audio, 401 North Broad St., Philadelphia, PA 19108

BYTE, 70 Main St., Peterborough, NH 03458

Computer Design, 11 Goldsmith St., Littleton, MA 01460

CQ, 14 Vanderventer Ave., Port Washington, L.I., NY 11050

Delco Electronics, 700 East Firmin, Kokomo, IN 46901

Dialight Corp., 203 Harrison Place, Brooklyn, NY 11237

EDN, 221 Columbus Ave., Boston, MA 02116

Electronics, 1221 Avenue of the Americas, New York, NY 10020

Electronic Servicing, 9221 Quivira Rd., P.O. Box 12901, Overland Park, KS 66212

Exar Integrated Systems, Inc., 750 Palomar Ave., Sunnyvale, CA 94086

Ham Radio, Greenville, NH 03048

Harris Semiconductor, Department 53-35, P.O. Box 883, Melbourne, FL 32901

Hewlett-Packard, 1501 Page Mill Rd., Palo Alto, CA 94304

Howard W. Sams & Co. Inc., 4300 West 62nd St., Indianapolis, IN 46206

IEEE Publications, 345 East 47th St., New York, NY 10017

Instruments & Control Systems, Chilton Way, Radnor, PA 19089

Kilobaud, Peterborough, NH 03458

McGraw-Hill Book Co., 1221 Avenue of the Americas, New York, NY 10020

Modern Electronics, 14 Vanderventer Ave., Port Washington, NY 11050

Motorola Semiconductor Products Inc., Box 20912, Phoenix, AZ 85036

Mullard Limited, Mullard House, Torrington Place, London WC1E 7HD, England

National Semiconductor Corp., 2900 Semiconductor Dr., Santa Clara, CA 95051

Optical Electronics Inc., P.O. Box 11140, Tucson, AZ 85734

Popular Science, 380 Madison Ave., New York, NY 10017

Precision Monolithics Inc., 1500 Space Park Dr., Santa Clara, CA 95050

QST, American Radio Relay League, 225 Main St., Newington, CT 06111

Radio Shack, 1100 One Tandy Center, Fort Worth, TX 76102

Raytheon Semiconductor, 350 Ellis St., Mountain View, CA 94042

RCA Solid State Division, Box 3200, Somerville, NJ 08876

Howard W. Sams & Co. Inc., 4300 West 62nd St., Indianapolis, IN 46206

73 Magazine, Peterborough, NH 03458

Siemens Corp., Components Group, 186 Wood Ave. South, Iselin, NJ 08830

Signetics Corp., 811 East Arques Ave., Sunnyvale, CA 94086

Siliconix Inc., 2201 Laurelwood Rd., Santa Clara, CA 95054

Sprague Electric Co., 479 Marshall St., North Adams, MA 01247

Teledyne Philbrick, Allied Drive at Route 128, Dedham, MA 02026

Teledyne Semiconductor, 1300 Terra Bella Ave., Mountain View, CA 94040

Texas Instruments Inc., P.O. Box 5012, Dallas, TX 75222

TRW Power Semiconductors, 14520 Aviation Blvd., Lawndale, CA 90260

Unitrode Corp., 580 Pleasant St., Watertown, MA 02172

Wireless World, Dorset House, Stamford St., London SE1 9LU, England

About the Author

John Markus is a professional writer residing in Sunnyvale, California. He serves as a special consultant to the McGraw-Hill Book Company, an organization he was associated with for 27 years before he struck out on his own as a writer and consultant. During that time he held many positions, including that of feature editor on *Electronics* magazine.

In this capacity he was responsible for many state-of-the-art reports in the field of electronics. One of these reports earned him the Jesse H. Neal Editorial Award for outstanding journalism.

He later served as technical director on the Technical Research Staff of the McGraw-Hill Book Company, applying electronic techniques to the mechanization of information publishing systems. His last assignment was as manager of information research for McGraw-Hill divisions on nontraditional publishing and information retrieval.

Mr. Markus is a senior member of the Institute of Electrical and Electronics Engineers. He is the author, coauthor, and editor of numerous books for McGraw-Hill, including *Electronics Dictionary*, Fourth Edition; *Television and Radio Repairing*, Second Edition; *How to Make More Money in Your TV Servicing Business; Sourcebook of Electronic Circuits; Electronic Circuits Manual*; and *Guidebook of Electronic Circuits*. He also is consulting editor and contributing editor to the McGraw-Hill *Dictionary of Scientific and Technical Terms*, and has contributed over 30 articles to the 15-volume McGraw-Hill *Encyclopedia of Science and Technology*, Fourth Edition.

ELECTRONICS PROJECTS
READY-REFERENCE

CHAPTER 1
Automotive Circuits

Includes capacitor-discharge, optoelectronic, and other types of electronic ignition, tachometers, dwell meters, idiot-light buzzer, audible turn signals, headlight reminders, mileage computer, cold-weather starting aids, wiper controls, oil-pressure and oil-level gages, solid-state regulators for alternators, overspeed warnings, battery-voltage monitor, and trailer-light interface. For auto theft devices, see Burglar Alarm chapter.

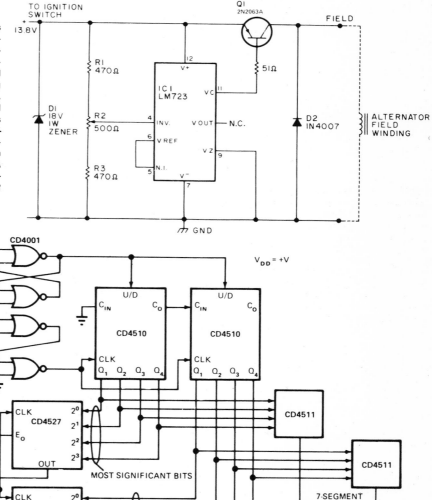

SOLID-STATE AUTO REGULATOR—Replaces and outperforms electromechanical charging-voltage regulator in autos using alternator systems. Prolongs battery life by preventing undercharging or overcharging of 12-V lead-acid battery. Uses LM723 connected as switching regulator for controlling alternator field current. R2 is adjusted to maintain 13.8-V fully charged voltage for standard auto battery. Article gives construction details and tells how to use external relay to maintain alternator charge-indicator function in cars having idiot light rather than charge-discharge ammeter. Q1 is 2N2063A (SK3009) 10-A PNP transistor.—W. J. Prudhomme, Build Your Own Car Regulator, *73 Magazine,* March 1977, p 160–162.

MILEAGE COMPUTER—Fuel consumption in miles per gallon is continuously updated on 2-digit LED display. Entire system using CMOS ICs can be built for less than $25 including gas-flow sensor and speed sensor, sources for which are given in article along with operational details. Circuit uses rate multiplier to produce output pulse train whose frequency is proportional to product of the two inputs. Output rate is time-averaged. Speed sensor, mounted in series with speedometer cable, feeds speed data to CD4527 rate multiplier as clock input. Gas-flow sensor, mounted in series with fuel line, feeds clock input of other rate multiplier.—G. J. Summers, Miles/Gallon Measurement Made Easy with CMOS Rate Multipliers, *EDN Magazine,* Jan. 20, 1976, p 61–63.

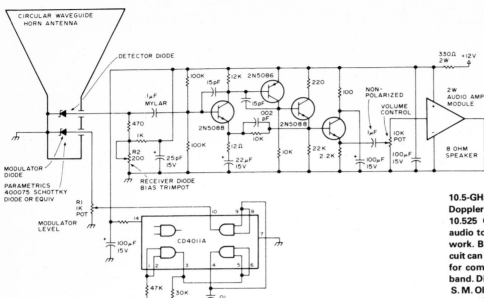

10.5-GHz RADAR DETECTOR—Picks up CW Doppler traffic radar signals in X-band region at 10.525 GHz and alerts speeding driver with audio tone. Article also tells how traffic radars work. By adding 10.5-GHz oscillator, same circuit can be used in 10.5-GHz amateur radio band for communicating with other cars using this band. Dimensioned diagram of horn is given.— S. M. Olberg, Mobile Smokey Detector, *73 Magazine*, Holiday issue 1976, p 32–35.

SPEED TRAP—Time required for auto to activate sensors placed measured distance apart on driveway or road is used to energize relay or alarm circuit when auto exceeds predetermined speed. If speed limit chosen is 15 mph, set detectors 22 feet apart for travel time of 1 s. Sensors can be photocells or air-actuated solenoids. For most applications, R1 can be 1-megohm pot. Transistor type is not critical. Values of R2 and C2 determine how long alarm sounds.—J. Sandler, 9 Projects under $9, *Modern Electronics,* Sept. 1978, p 35–39.

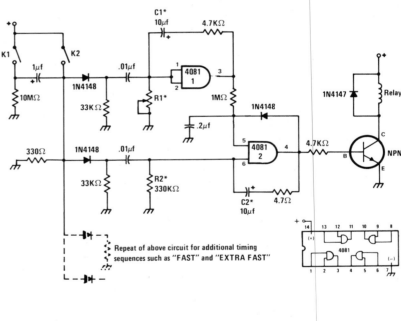

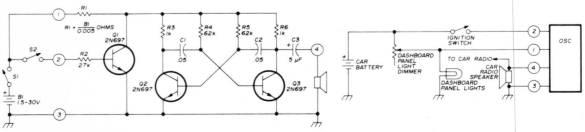

HEADLIGHT REMINDER—Uses basic oscillator consisting of Q2 and Q3 arranged as collector-coupled astable MVBR. Power is taken from collector of Q1 which acts as switch for Q2 and Q3. With S1 closed and S2 open, oscillator operates. Closing S2 saturates Q1 and stops oscillator. When used as headlight reminder for negative-ground car, B1 is omitted and power for oscillator is taken from dashboard panel lights since they come on simultaneously with either parking lights or headlights. If ignition key is turned on, Q1 saturates and disables Q2-Q3. With ignition off but lights on, Q1 is cut off and oscillator receives power. Audio output may be connected directly to high side of voice coil of car radio loudspeaker without affecting operation of radio. Almost any NPN transistors can be used. Changing values of R4 and R5 changes frequency of reminder tone.—H. F. Batie, Versatile Audio Oscillator, *Ham Radio,* Jan. 1976, p 72–74.

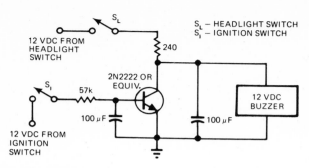

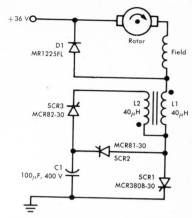

HEADLIGHTS-ON ALARM—Designed for cars in which headlight switch is nongrounding type, providing 12 V when closed. When both light and ignition switches are closed, transistor is saturated and there is no voltage drop across it to drive buzzer. If ignition switch is open while lights are on, transistor bias is removed so transistor is effectively open and full 12 V is applied to buzzer through 240-ohm resistor until lights are turned off.—R. E. Hartzell, Jr., Detector Warns You When Headlights Are Left On, *EDN Magazine*, Nov. 20, 1975, p 160.

ELECTRIC-VEHICLE CONTROL—SCR1 is used in combination with Jones chopper to provide smooth acceleration of golf cart or other electric vehicle operating from 36-V on-board storage battery. Normal running current of 2-hp 36-V series-wound DC motor is 60 A, with up to 300 A required for starting vehicle up hill. Chopper and its control maintain high average motor current while limiting peak current by increasing chopping frequency from normal 125 Hz to as high as 500 Hz when high torque is required.—T. Malarkey, You Need Precision SCR Chopper Control, *New Motorola Semiconductors for Industry*, Motorola, Phoenix, AZ, Vol. 2, No. 1, 1975.

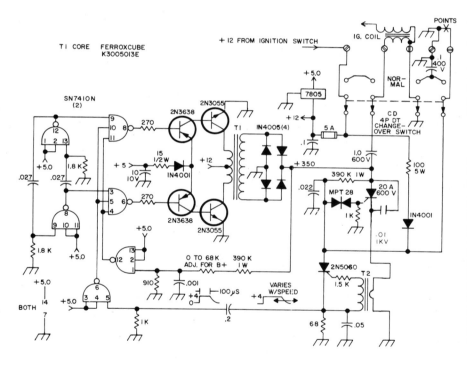

CD IGNITION—Uses master oscillator—power amplifier type of DC/DC converter in which two sections of triple 3-input NAND gate serve as 10-kHz square-wave MVBR feeding class B PNP/NPN power amplifier through two-gate driver. Remaining two gates are used as logic inverters. Secondary of T1 has 15.24 meters of No. 26 in six bank windings, with 20 turns No. 14 added and center-tapped for primary. T2 is unshielded iron-core RF choke, 30–100 μH, with several turns wound over it for secondary. When main 20-A SCR fires, T2 develops oscillation burst for firing sensitive gate-latching SCR. Storage capacitor energy is then dumped into ignition coil primary through power SCR.—K. W. Robbins, CD Ignition System, *73 Magazine*, May 1974, p 17 and 19.

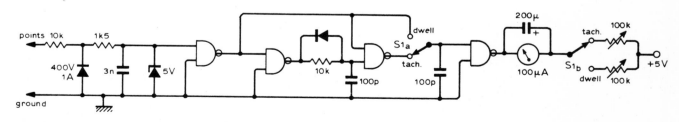

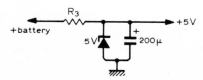

TACH/DWELL METER—Built around SN7402 NOR-gate IC. Requires no internal battery; required 5 V is obtained by using 50 ohms for R_3 in zener circuit shown if car battery is 6 V, and 300 ohms if 12 V. Article gives calibration procedure for engines having 4, 6, and 8 cylinders; select maximum rpm to be indicated, multiply by number of cylinders, then divide by 120 to get frequency in Hz.—N. Parron, Tach-Dwell Meter, *Wireless World*, Sept. 1975, p 413.

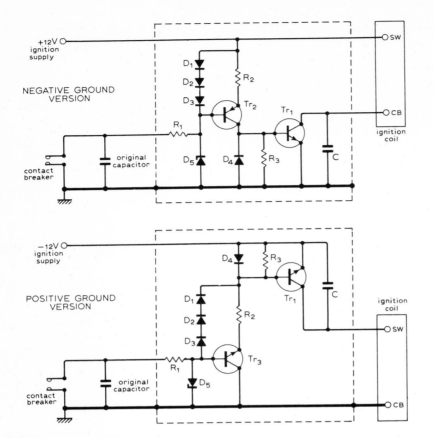

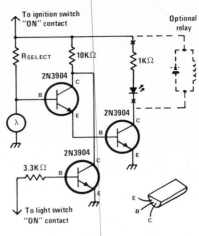

HEADLIGHT REMINDER—Photocell energizes circuit at twilight to remind motorist that lights should be turned on. Indicator can be LED connected as shown or relay turning on buzzer for more positive signal. Circuit can be made automatic by connecting relay contacts in parallel with light switch, provided delay circuit is added to prevent oncoming headlights from killing circuit. Mount photocell in location where it is unaffected by other lights inside or outside car.—J. Sandler, 9 Projects under $9, *Modern Electronics*, Sept. 1978, p 35–39.

TRANSISTORIZED BREAKER POINTS—Uses Texas Instruments BUY23/23A high-voltage transistors that can easily withstand voltages up to about 300 V existing across breaker points of distributor in modern car. Circuit serves as electronic switch that isolates points from heavy interrupt current and high-voltage back-swing of ignition coil, thereby almost completely eliminating wear on points. Values are: Tr_2 2N3789; Tr_3 (for positive ground version) 2N3055; D_1-D_4 1N4001; D_5 18-V 400-mW zener; R_1 56 ohms; R_2 1.2 ohms; R_3 10 ohms; C 600 VDC same size as points capacitor. Article covers installation procedure.—G. F. Nudd, Transistor-Aided Ignition, *Wireless World*, April 1975, p 191.

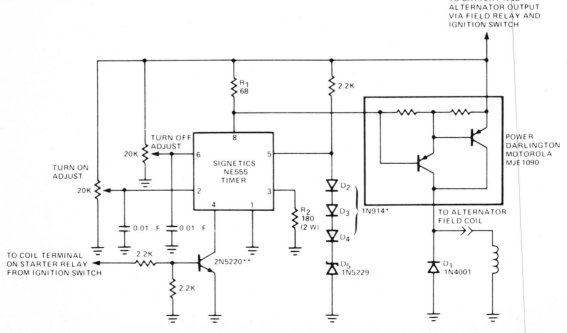

VOLTAGE REGULATOR—Timer and power Darlington form simple automobile voltage regulator. When battery voltage drops below 14.4 V, timer is turned on and Darlington pair conducts. Separate adjustments are provided for preset turn-on and turnoff voltages.—"Signetics Analog Data Manual," Signetics, Sunnyvale, CA, 1977, p 731.

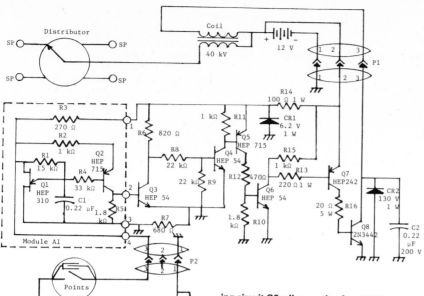

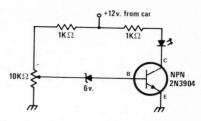

BATTERY MONITOR—Basic circuit energizes LED when battery voltage drops to level set by 10K pot. Any number of additional circuits can be added, for reading battery voltage in 1-V steps or even steps as small as 0.1 V. Circuit supplements idiot light that replaces ammeter in most modern cars. LED type is not critical.—J. Sandler, 9 Projects under $9, *Modern Electronics,* Sept. 1978, p 35–39.

COLD-WEATHER IGNITION—Multispark electronic ignition improves cold-weather starting ability of engines in arctic environment by providing more than one spark per combustion cycle. Circuit uses UJT triangle-wave generator Q1, emitter-follower isolator Q2, wave-shaping Schmitt trigger Q3-Q4, three stages of square-wave amplification Q5-Q7, and output switching circuit Q8, all operating from 12-V negative-ground supply. 6.2-V zener provides regulated voltage for UJT and Schmitt trigger. Initial 20,000- to 40,000-V ignition spark produced by opening of breaker points is followed by continuous series of sparks at rate of about 200 per second as long as points stay open.—D. E. Stinchcomb, Multi-Spark Electronic Ignition for Engine Starting in Arctic Environment, *Proceedings of the IEEE 1975 Region Six Conference,* May 1975, p 224–225.

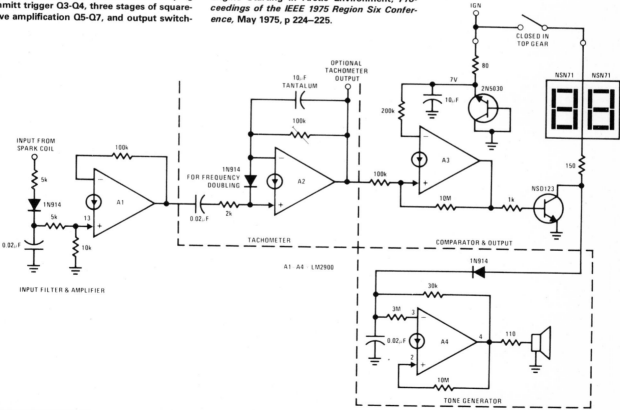

HIGH-SPEED WARNING—Audible alarm tone generator drives warning loudspeaker to supplement 2-digit speed display that can be set to trip when vehicle speed exceeds 55-mph legal limit. Engine speed signal is taken from primary of spark coil. Switch in transmission activates circuit only when car is in high gear. All functions are performed by sections of LM2900 quad Norton opamp. A1 amplifies and regulates spark-coil signal. A2 converts signal frequency to voltage proportional to engine speed. A3 compares speed voltage with reference voltage and turns on output transistor at set speed. A4 generates audible tone. Circuit components must be adjusted for number of cylinders, gear and axle ratios, tire size, etc. 10-μF capacitor connected to A3 can be increased to prevent triggering of alarm when increasing speed momentarily while passing another car.—"Linear Applications, Vol. 2," National Semiconductor, Santa Clara, CA, 1976, LB-33.

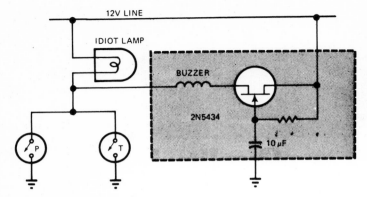

BUZZER FOR IDIOT LIGHT—Provides audible supplement to engine-monitoring indicator lamps that are often difficult to see in daylight. Uses 2N5434 JFET to provide delay of about 7 s each time ignition switch is turned on, to allow for peaceful starting of car and normal buildup of oil pressure when lamp is monitoring oil-pressure and engine-temperature sensors. Entire circuit can be mounted inside plastic housing of unused or disconnected dashboard warning buzzer in late-model car.—P. Clower, Audio Assist Gives "Idiot Lights" the "Buzz," *EDN Magazine,* June 20, 1976, p 126.

OIL-PRESSURE DISPLAY—Red, yellow, and green LEDs give positive indication of oil pressure level on electronic gage console developed for motorcycle. Transducer converts oil pressure to variable resistance R_T which in turn varies bias on transistors. LEDs have different forward voltages at which they light, so proper selection of bias resistors ensures that only one LED is on at a time to give desired indication of oil pressure.—J. D. Wiley, Instrument Console Features Digital Displays and Built-In Combo Lock, *EDN Magazine,* Aug. 5, 1975, p 38–43.

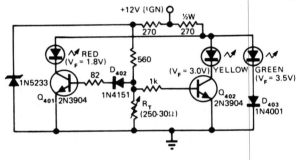

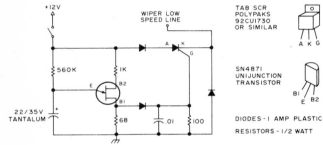

WIPER CONTROL—Operates wipers automatically at intervals, as required for very light rain or mist. Changing 560K resistor to 500K pot in series with 100K fixed resistor gives variable control of interval.—Circuits, *73 Magazine,* July 1977, p 34.

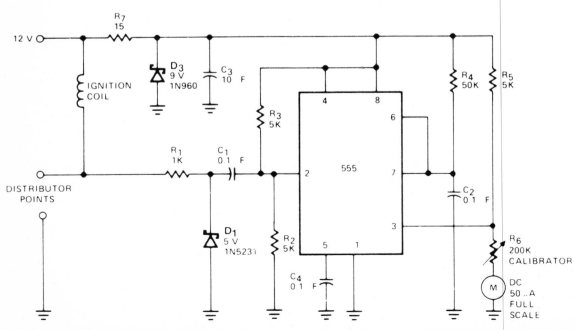

DISTRIBUTOR-POINT TACHOMETER—555 timer receives its input pulses from distributor points of car. When timer output (pin 3) is high, meter receives calibrated current through R_6. When IC times out, meter current stops for remainder of duty cycle. Integration of variable duty cycle by meter movement serves to provide visible indication of engine speed.—"Signetics Analog Data Manual," Signetics, Sunnyvale, CA, 1977, p 724–725.

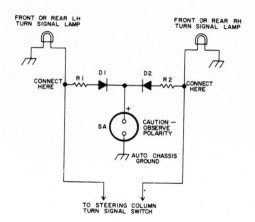

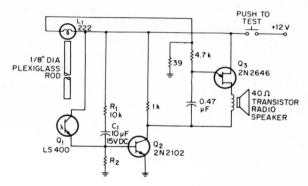

AUDIBLE TURN SIGNAL—Gives 3500-Hz audible tone each time turn-signal light flashes on, to warn driver that signal has not been turned off when making less than right-angle turns. Schematic shown is for 12-V negative-ground systems. For 6-V negative-ground systems, cut values of R1 and R2 about in half. For positive-ground systems, reverse connections to diodes and Sonalert. R1 and R2 are 2.7K 0.5 W. D1 and D2 can be any general-purpose small-current silicon diode. SA is Mallory SC1.5 Sonalert.—A. Goodwin, Turn Signal Reminder, *73 Magazine,* Holiday issue 1976, p 166.

OIL-LEVEL GAGE—Permits checking crankcase oil level from driver's seat. Sensor consists of light-conducting Plexiglas rod attached to dipstick, with lamp L_1 at top of rod and phototransistor Q_1 mounted at add-oil mark on dipstick, about ½ inch below bottom of rod. At normal oil level, oil attenuates light between Q_1 and bottom end of rod, making phototransistor resistance high. Pushing test switch makes C_1 charge and saturate Q_2 long enough to activate UJT AF oscillator Q_3 and give short tone verifying that lamp is not burned out and gage is working. When oil is low, enough light reaches Q_1 to keep Q_2 saturated after C_1 charges, giving continuous tone as long as switch is pushed.— L. Svelund, Electronic Dipstick, *EEE Magazine,* Nov. 1970, p 101.

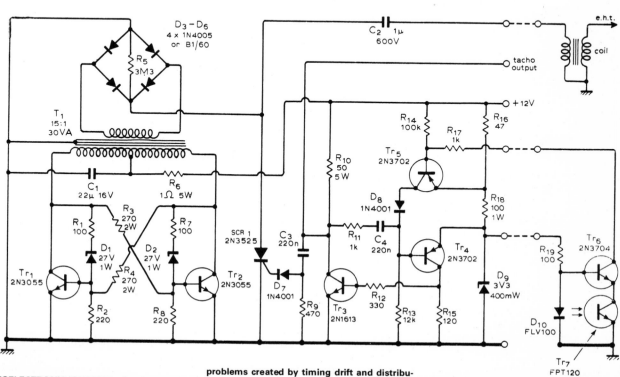

OPTOELECTRONIC IGNITION—Combination of low-cost point-source LED and high-sensitivity phototransistor forms optical sensor for position of cam in distributor. Technique eliminates problems created by timing drift and distributor-shaft play. Sensor head is small enough to fit most distributors. Article gives dimensioned drawings for shutter design and sensor mounting, and describes operation of associated capacitor-discharge electronic ignition circuit in detail. Leads to sensor do not require shielding.—H. Maidment, Optical Sensor Ignition System, *Wireless World,* Nov. 1975, p 533–537.

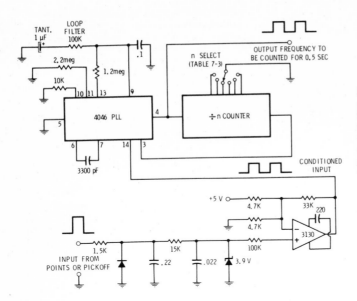

DIGITAL TACHOMETER—Pulses from auto engine points or other pickoff are filtered before feed to 3130 CMOS opamp used as comparator to complete conditioning of input. Pulses are then fed through 4046 PLL to divide-by-N counter that is set for number of cylinders in engine (60 for four cylinders, 45 for six, and 30 for eight). Output frequency is then counted for 0.5 s to get engine or shaft speed in rpm.—D. Lancaster, "CMOS Cookbook," Howard W. Sams, Indianapolis, IN, 1977, p 366–367.

REGULATOR FOR ALTERNATOR—Simple and effective solid-state replacement for auto voltage regulator can be used with alternator in almost any negative-ground system. Circuit acts as switch supplying either full or no voltage to field winding of alternator. When battery is below 13 V, zener D1 does not conduct, Q1 is off, Q2 is on, and full battery voltage is applied to alternator field so it puts out full voltage to battery for charging. When battery reaches 13.6 V, Q1 turns on, Q2 turns off, alternator output is reduced to zero, and battery gets no charging

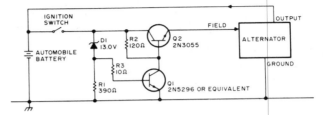

current. Circuit can also be used with wind-driven alternator systems.—P. S. Smith, $22 for a Regulator? Never!, *73 Magazine,* Holiday issue 1976, p 103.

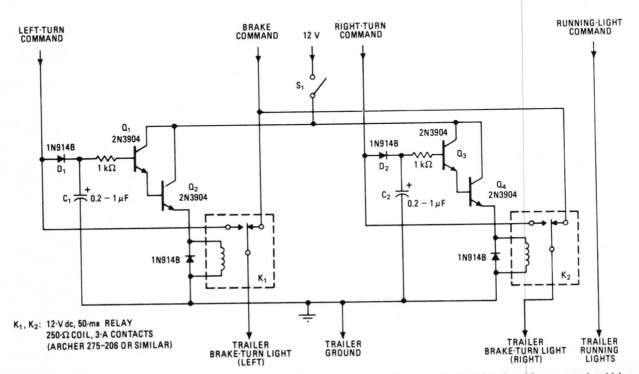

K_1, K_2: 12-V dc, 50-ma RELAY
250-Ω COIL, 3-A CONTACTS
(ARCHER 275-206 OR SIMILAR)

AUTO-TRAILER INTERFACE FOR LIGHTS—Low-cost transistors and two relays combine brake-light and turn-indicator signals on common bus to ensure that trailer lights respond to both commands. C_1 and C_2 charge to peak amplitude of turn signal, which flashes about 2 times per second. Values are selected to hold relay closed between flash intervals; if capacitance is too large, brake signal cannot immediately activate trailer lights after turn signal is canceled. Developed for new cars in which separate turn and brake signals are required for safety.—M. E. Gilmore and C. W. Snipes, Darlington-Switched Relays Link Car and Trailer Signal Lights, *Electronics,* Aug. 18, 1977, p 116.

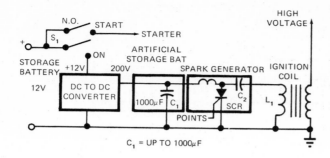

CAPACITOR SERVES AS IGNITION BATTERY— Developed for use with capacitor-discharge ignition systems to provide independent voltage source for ignition when starting car in very cold weather. Before attempting to start car, S_1 is set to ON position for energizing DC-to-DC converter for charging C_1 with DC voltage between 200 and 400 V. Starter is now engaged. If voltage of storage battery drops as starter slowly turns engine over, C_1 still represents equivalent of fully charged 12-V storage battery that is capable of driving ignition system for almost a minute.—W. Stalzer, Capacitor Provides Artificial Battery for Ignition Systems, *EDN Magazine,* Nov. 15, 1972, p 48.

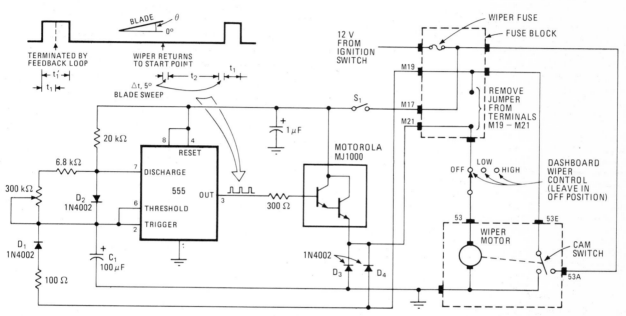

WIPER-DELAY CONTROL—555 timer provides selectable delay time between sweeps of wiper blades driven by motor in negative-ground system. Article also gives circuit modification for positive-ground autos. Delay time can be varied between 0 and 22 s. Timer uses feedback signal from cam-operated switch of motor to synchronize delay time with position of wiper blades.— J. Okolowicz, Synchronous Timing Loop Controls Windshield Wiper Delay, *Electronics,* Nov. 24, 1977, p 115 and 117.

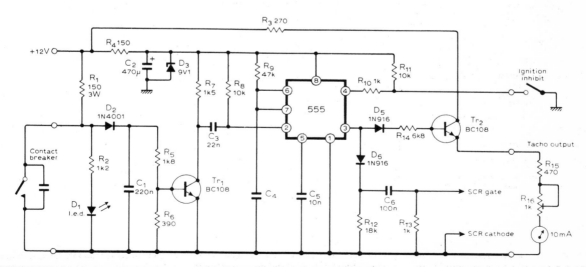

RPM-LIMIT ALARM—Used with capacitor-discharge ignition system to provide tachometer output along with engine speed control signal. When breaker contacts open, C_1 charges and turns Tr_1 on, triggering 555 timer used in mono MVBR mode. Resulting positive pulse from 555 fires control SCR through D_6 and C_6. When contacts close, D_2 isolates C_1 to reduce effect of contact bounce. With values shown, for speed limit between 8000 and 9000 rpm, use 0.068 μF for C_4 with four-cylinder engine, 0.047 μF for six cylinders, and 0.033 μF for eight cylinders. LED across breaker contacts can be used for setting static timing.—K. Wevill, Trigger Circuit for C.D.I. Systems, *Wireless World,* Jan. 1978, p 58.

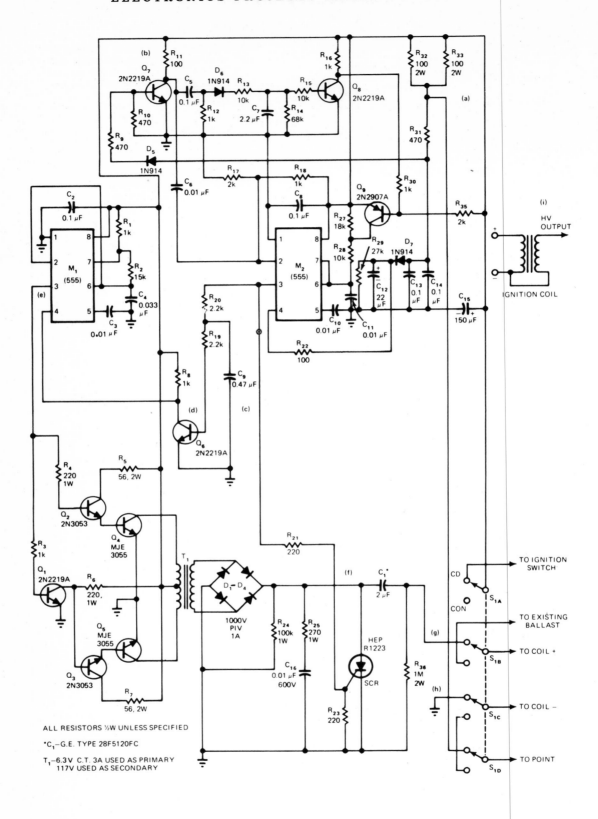

ALL RESISTORS ½W UNLESS SPECIFIED

*C₁–G.E. TYPE 28F5120FC

T₁–6.3V C.T. 3A USED AS PRIMARY
117V USED AS SECONDARY

LOW-EMISSION CD—Solid-state capacitor-discharge ignition system improves combustion efficiency by increasing spark duration. For 8-cylinder engine, normal CD system range of 180 to 300 μs is increased to 600 μs below 4000 rpm. Oscillation discharge across ignition coil primary lasts for two cycles here, but above 4000 rpm the discharge lasts for one cycle or 300 μs because at higher speeds the power cycle has shorter times. Circuit uses 555 timer M_1 as 2-kHz oscillator, with Q_1-Q_3 providing drive to Q_4-Q_5 and T_1 for converting battery voltage to about 400 VDC at output of bridge rectifier. When distributor points open, Q_7 turns on and triggers M_2 connected as mono that provides gate drive pulses for SCR. Article describes operation of circuit in detail and gives waveforms at points a-i.—C. C. Lo, CD Ignition System Produces Low Engine Emissions, *EDN Magazine*, May 20, 1976, p 94, 96, and 98.

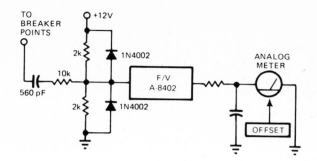

TACHOMETER—Intech/Function Modules A-8402 operating in frequency-to-voltage converter mode serves as automotive tachometer having inherent linearity and ease of calibration. Converter operates asynchronously, which does not affect accuracy when driving analog meter.—P. Pinter and D. Timm, Voltage-to-Frequency Converters—IC Versions Perform Accurate Data Conversion (and Much More) at Low Cost, *EDN Magazine,* Sept. 5, 1977, p 153–157.

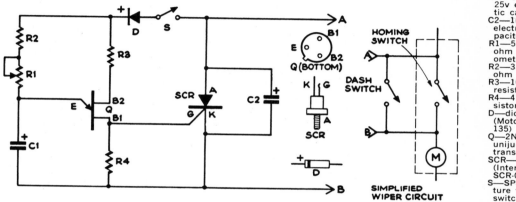

C1—50ufd @ 25v electrolytic capacitor
C2—1ufd @25v electrolytic capacitor
R1—50,000-ohm potentiometer
R2—33,000-ohm resistor
R3—100-ohm resistor
R4—47-ohm resistor
D—diode (Moto. HEP 135)
Q—2N1671B unijunction transistor
SCR—SCR (Inter. Rect. SCR-03)
S—SPST miniature toggle switch

TIMER FOR WIPER—Provides automatic one-shot swipes at preselected intervals from 2 to 30 s for handling mist, drizzle, or splash from wet road. Circuit shorts out homing switch inside windshield-wiper motor, which is usually in parallel with slow-speed contacts of wiper dashboard switch. With wiper switch off and ignition on, short two switch terminals at a time to find pins that start wiper. When blades begin moving, remove jumper; blades should then finish sweep and shut off. It is these terminals of switch that are connected to points A and B of control circuit.—V. Mele, Mist Switch—It's for Your Windshield Wipers, *Popular Science,* Aug. 1973, p 110.

CHAPTER **2**
Battery-Charging Circuits

Includes constant-voltage, constant-current, and trickle chargers operating from AC line, solar cells, or auto battery. Some circuits have automatic charge-rate control, automatic start-up, automatic shutoff, and low-charge indicator.

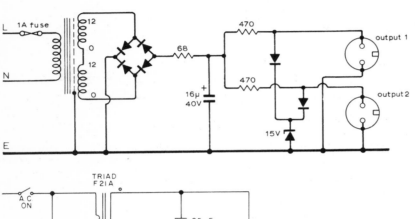

9.6 V AT 20 mA—Developed to charge 200-mAh nickel-cadmium batteries for two transceivers simultaneously. Batteries will be fully charged in 14 hours, using correct 20-mA charging rate. Zener diode ensures that voltage cannot exceed safe value if battery is accidentally disconnected while under charge. Diode types are not critical.—D. A. Tong, A Pocket V.H.F. Transceiver, *Wireless World,* Aug. 1974, p 293–298.

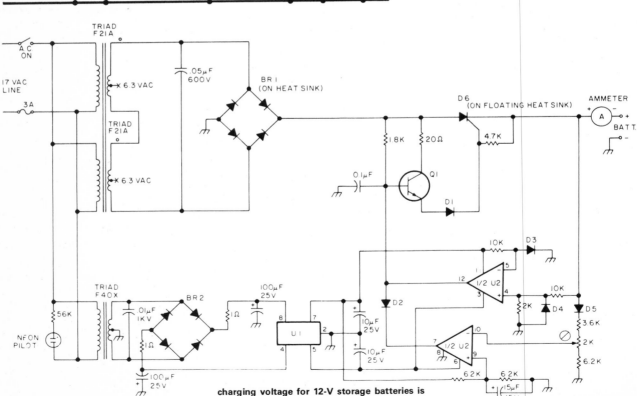

ADJUSTABLE FINISH-CHARGE—Uses National LM319D dual comparator U2 to sense end-of-charge battery voltage and provide protection against shorted or reversed charger leads. Final charging voltage for 12-V storage batteries is adjustable with 2K trimpot. Separate ±15 V supply using Raytheon RC4195NB regulator U1 is provided for U2. D1-D5 are 1N4002 or HEP-R0051. D6 is 2N682 or HEP-R1471. BR1 is Motorola MDA980-2 or HEP-R0876 12-A bridge. BR2 is Varo VE27 1-A bridge. Q1 is 2N3641 or HEP-S0015.—H. Olson, Battery Chargers Exposed, *73 Magazine,* Nov. 1976, p 98–100 and 102–104.

NICKEL-CADMIUM CELL CHARGER—Charges four size D cells in series at constant current, with automatic voltage limiting. BC301 transistor acts as current source, with base voltage stabilized at about 3 V by two LEDs that also serve to indicate charge condition. Other transistor provides voltage limiting when voltage across cells approaches that of 1K branch of voltage divider. Values shown give 260-mA charge initially, dropping to 200 mA when V_c reaches 5 V and decreasing almost to 0 when V_c reaches 6.5 V.—N. H. Sabah, Battery Charger, *Wireless World,* Nov. 1975, p 520.

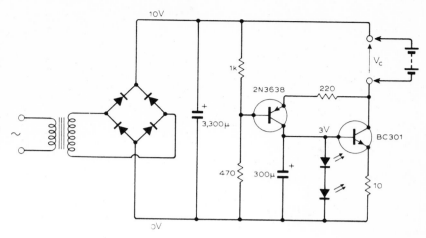

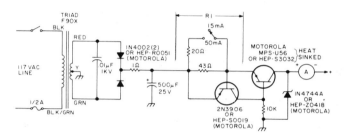

12-V FOR NICADS—Produces constant current with simple transistor circuit, adjustable to 15 or 50 mA with switch and R1. Zener limits voltage at end of charge. Developed for charging 10-cell pack having nominal 12.5 V, as used in many transceivers.—H. Olson, Battery Chargers Exposed, *73 Magazine,* Nov. 1976, p 98–100 and 102–104.

CHARGING SILVER-ZINC CELLS—Used for initial charging and subsequent rechargings of sealed dry-charged lightweight cells developed for use in missiles, torpedoes, and space applications. Article covers procedure for filling cell with potassium hydroxide electrolyte before placing in use (cells are dry-charged at factory and have shelf life of 5 or more years in that condition). Charge current should be 7 to 10% of rated cell discharge capacity; thus, for Yardney HR-5 cell with rated discharge of 5 A, charge at 350 to 500 mA. Stop charging when cell voltage

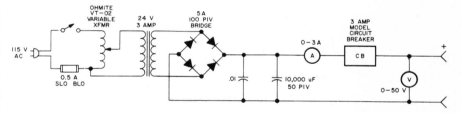

reaches 2.05 V. If used only for battery charging, large filter capacitor can be omitted.—S. Kelly, Will Silver-Zinc Replace the Nicad?, *73 Magazine,* Holiday issue 1976, p 204–205.

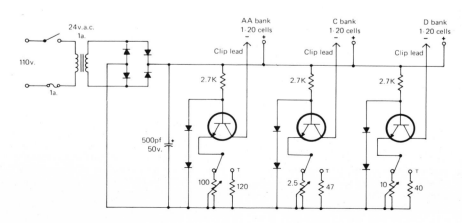

BULK NICAD CHARGER—Can handle up to 20 AA cells, 20 C cells, and 20 D cells simultaneously, with charging rate determined separately for each type. Single transformer and full-wave rectifier feed about 24 VDC to three separate regulators. AA-cell regulator uses 100-ohm resistor to vary charge rate from 6 mA to above 45 mA. C-cell charge-rate range is 24 to 125 mA, and D-cell range is 60 to 150 mA. Batteries of each type should be about same state of discharge. Batteries are recharged in series to avoid need for separate regulator with each cell. Trickle-charge switches cut charge rates to about 2% of rated normal charge (5 mA for 500-mAh AA cells). Transistors are 2N4896 or equivalent. Use heatsinks. All diodes are 1N4002.—J. J. Schultz, A Bulk Ni-Cad Recharger, *CQ,* Dec. 1977, p 35–36 and 111.

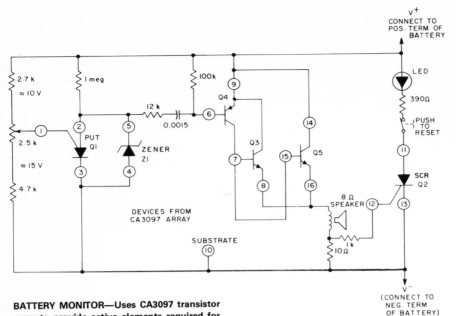

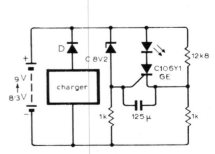

SOLAR-POWER BACKUP—If solar-cell voltage drops 0.2 V below battery voltage, circuit is powered by storage cell feeding through forward-biased OA90 or equivalent germanium diode. When solar-cell voltage exceeds that of battery, battery is charged by approximately constant reverse leakage current through diode. Battery can be manganese-alkaline type or zinc-silver oxide watch-type cell.—M. Hadley, Automatic Micropower Battery Charger, *Wireless World,* Dec. 1977, p 80.

BATTERY MONITOR—Uses CA3097 transistor array to provide active elements required for driving indicators serving as aural and visual warnings of low charge on nicad battery. LED remains on until circuit is reset with pushbutton switch.—"Circuit Ideas for RCA Linear ICs," RCA Solid State Division, Somerville, NJ, 1977, p 9.

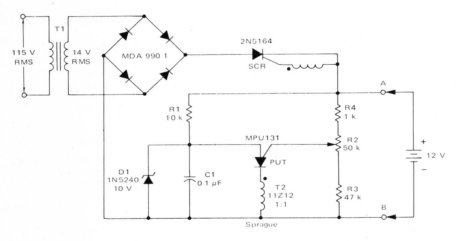

12 V AT 8 A—Charging circuit for lead-acid storage batteries is not damaged by short-circuits or by connecting with wrong battery polarity. Battery provides current for charging C1 in PUT relaxation oscillator. When PUT is fired by C1, SCR is turned on and applies charging current to battery. Battery voltage increases slightly during charge, increasing peak point voltage of PUT and making C1 charge to slightly higher voltage. When C1 voltage reaches that of zener D1, oscillator stops and charging ceases. R2 sets maximum battery voltage between 10 and 14 V during charge.—R. J. Haver and B. C. Shiner, "Theory, Characteristics and Applications of the Programmable Unijunction Transistor," Motorola, Phoenix, AZ, 1974, AN-527, p 10.

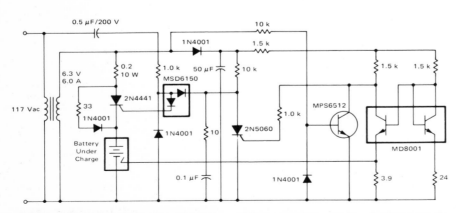

THIRD ELECTRODE SENSES FULL CHARGE—Circuit is suitable only for special nickel-cadmium batteries in which third electrode has been incorporated for use as end-of-charge indicator. Voltage change at third electrode is sufficient to provide reliable shutoff signal for charger under all conditions of temperature and cell variations.—D. A. Zinder, "Fast Charging Systems for Ni-Cd Batteries," Motorola, Phoenix, AZ, 1974, AN-447, p 7.

LED VOLTAGE INDICATOR—Circuit shown uses LED to indicate, by lighting up, that battery has been charged to desired level of 9 V. Circuit can be modified for other charging voltages. Silicon switching transistor can be used in place of more costly thyristor.—P. R. Chetty, Low Battery Voltage Indication, *Wireless World,* April 1975, p 175.

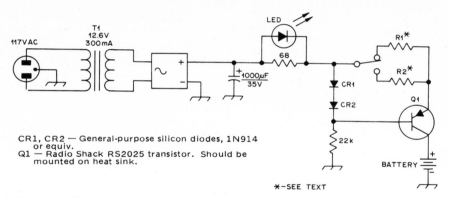

CR1, CR2 — General-purpose silicon diodes, 1N914 or equiv.
Q1 — Radio Shack RS2025 transistor. Should be mounted on heat sink.

*—SEE TEXT

NICAD CHARGER—Switch gives choice of two constant-current charge rates. With 10 ohms for R1, rate is 60 mA, while 200 ohms for R2 gives 3 mA. Silicon diodes CR1 and CR2 have combined voltage drop of 1.2 V and emitter-base junction of Q1 has 0.6-V drop, for net drop of 0.6 V across R1 or R2. Dividing 0.6 by desired charge rate in amperes gives resistance value.—M. Alterman, A Constant-Current Charger for Nicad Batteries, *QST*, March 1977, p 49.

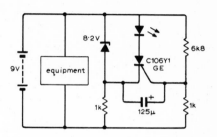

LED INDICATES LOW VOLTAGE—LED lights when output of 9-V rechargeable battery drops below minimum acceptable value of 8.3 V, to indicate need for recharging. Can also be used with transistor radio battery to indicate need for replacement. Zener is BZY85 C8V2 rated at 400 mW, with avalanche point at 7.7 V because of low current drawn by circuit. LED can be Hewlett-Packard 5082-4440.—P. C. Parsonage, Low-Battery Voltage Indicator, *Wireless World*, Jan. 1973, p 31.

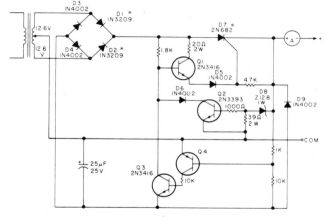

12-V AUTOMATIC—Circuit of Heathkit GP-21 automatic charger is self-controlling (Q1 and Q2) and provides protection against shorted or reversed battery leads (Q3 and Q4). Zener D8 is not standard value, so may be obtainable only in Heathkits. D1, D2, and D7 should all be on one heatsink.—H. Olson, Battery Chargers Exposed, *73 Magazine*, Nov. 1976, p 98—100 and 102—104.

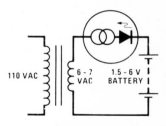

LED TRICKLE CHARGER—Constant-current characteristic of National NSL4944 LED is used to advantage in simple half-wave charger for batteries up to 6 V.—"Linear Applications, Vol. 2," National Semiconductor, Santa Clara, CA, 1976, AN-153, p 2.

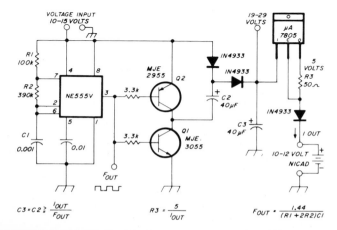

$$C3 = C2 = \frac{I_{OUT}}{F_{OUT}}$$

$$R3 = \frac{5}{I_{OUT}}$$

$$F_{OUT} = \frac{1.44}{(R1 + 2R2)C1}$$

NICAD CHARGER FOR AUTO—Voltage doubler provides at least 20 V from 12-V auto battery, for constant-current charging of 12-V nicads, using NE555 timer and two power transistors. Doubled voltage drives source current into three-terminal current regulator. Switching frequency of NE555 as MVBR is 1.4 kHz. Charging current is set at 50 mA for charging ten 500-mAh nicads.—G. Hinkle, Constant-Current Battery Charger for Portable Operation, *Ham Radio*, April 1978, p 34—36.

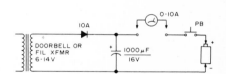

NICAD ZAPPER—Simple circuit often restores dead or defective nicad battery by applying DC overvoltage at current up to 10 A for about 3 s. Longer treatment may overheat battery and make it explode.—Circuits, *73 Magazine*, July 1977, p 35.

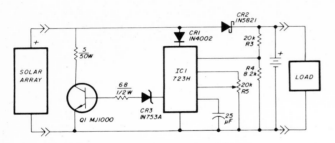

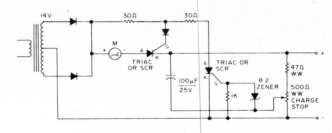

SOLAR-POWER OVERCHARGE PROTECTION— Voltage regulator is connected across solar-cell array as shown to prevent damage to storage battery by overcharging. Series diode prevents array from discharging battery during hours of darkness. Regulator does not draw power from battery, except for very low current used for voltage sampling. Battery can be lead-calcium, gelled-electrolyte, or telephone-type wet cells. For repeater application described, two Globe Union GC12200 40-Ah gelled-electrolyte batteries were used to provide transmit current of 1.07 A and idle current of 12 mA.—T. Handel and P. Beauchamp, Solar-Powered Repeater Design, *Ham Radio,* Dec. 1978, p 28–33.

AUTOMATIC SHUTOFF—Prevents overcharging and dryout of battery under charge by shutting off automatically when battery reaches full-charge voltage. Accepts wide range of batteries. Choose rectifying diodes and triacs or SCRs to handle maximum charging current desired. For initial adjustment, connect fully charged battery and adjust charge-stop pot until ammeter just drops to zero.—Circuits, *73 Magazine,* July 1977, p 34.

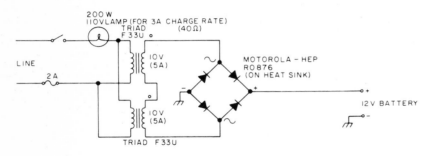

BASIC 12-V CHARGER—Uses 200-W lamp as current-limiting resistor in transformer primary circuit. Serves in place of older types of chargers using copper-oxide or tungar-bulb rectifiers.— H. Olson, Battery Chargers Exposed, *73 Magazine,* Nov. 1976, p 98–100 and 102–104.

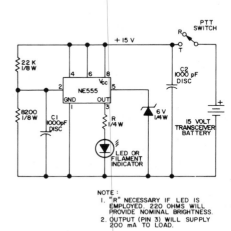

NOTE:
1. "R" NECESSARY IF LED IS EMPLOYED. 220 OHMS WILL PROVIDE NOMINAL BRIGHTNESS.
2. OUTPUT (PIN 3) WILL SUPPLY 200 mA TO LOAD.

NICAD MONITOR—Uses two comparators, flip-flop, and power stage all in single NE555 IC. When battery voltage drops below 12-V threshold set by R1 and R2 for 15-V transceiver battery, one comparator sets flip-flop and makes output at pin 3 go high. IC then supplies up to 200 mA to LED or other indicator. For other battery voltage value, set firing point to about three-fourths of fully charged voltage. Since battery voltage will show biggest drop when transmitting, connect monitor across transmit supply only so as to minimize battery drain.— A. Woerner, Ni-Cad Lifesaver, *73 Magazine,* Nov. 1973, p 35–36.

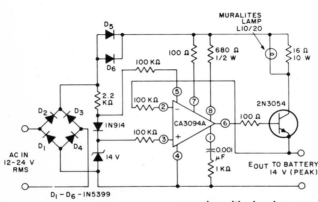

14-V MAXIMUM—Circuit accurately limits peak output voltage to 14 V, as established by zener connected between terminals 3 and 4 of CA3094A programmable opamp. Lamp brightness varies with charging current. Reference voltage supply does not drain battery when power supply is disconnected.—"Circuit Ideas for RCA Linear ICs," RCA Solid State Division, Somerville, NJ, 1977, p 19.

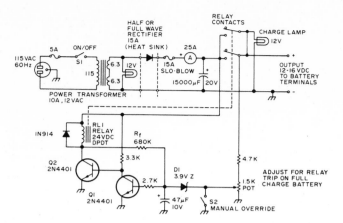

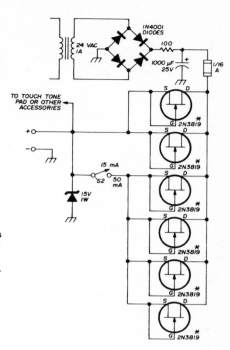

AUTOMATIC SHUTOFF—Charger automatically turns itself off when 12-V auto storage battery is fully charged. Setting of 1.5K pot determines battery voltage at which zener D1 conducts, turning on Q2 and pulling in relay that disconnects charger. If battery voltage drops below threshold, relay automatically connects charger again. S2 is closed to bypass automatic control when charger itself is to be used as power supply.—G. Hinkle, The Smart Charger, *73 Magazine,* Holiday issue 1976, p 110–111.

NICAD CHARGER—Pot is adjusted to provide 10% above rated voltage (normal full-charge voltage) while keeping charging current below 25% of maximum. For 10-V 1-Ah battery, set voltage at 11 V and current below 250 mA.—G. E. Zook, F.M., *CQ,* Feb. 1973, p 35–37.

NICAD CHARGER—Developed for recharging small nickel-cadmium batteries used in hand-held FM transceivers. Field-effect transistors serve as constant-current sources when gate is shorted to source. Practically any N-channel JFET having drain-to-source current of 8–15 mA will work. FETs shown were measured individually and grouped to give desired choice of 15- or 50-mA charging currents.—G. K. Shubert, FET-Controlled Charger for Small Nicad Batteries, *Ham Radio,* Aug. 1975, p 46–47.

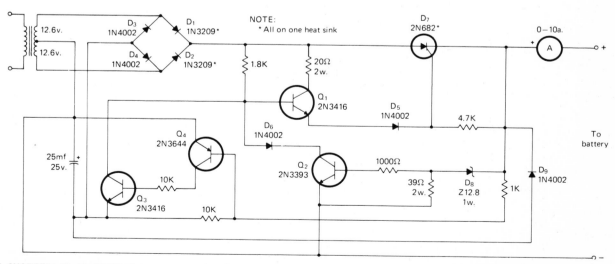

12-V CHARGER—Heath GP-21 charger uses SCR as switch to connect and disconnect battery at 120-Hz rate. Voltage at anode of SCR D_7 goes positive each half-cycle, putting forward bias on base of Q_1 through 1.8K resistor so Q_1 passes current through D_5 to gate of D_7 to turn it on for part of half-cycle and charge battery. D_7 stays on until voltage across it drops to zero. When battery has charged to 13.4 V, charging stops automatically. Rest of circuit protects against battery polarity reversal and accidental shorting of output leads. Special 12.8-V zener can be replaced by selected 1N4742 and forward-biased 1N4002.—H. Olson, We Don't Charge Nothin' but Batteries!, *CQ,* Feb. 1976, p 25–28 and 69.

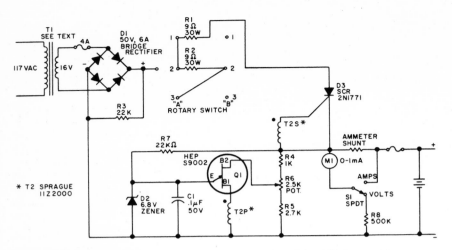

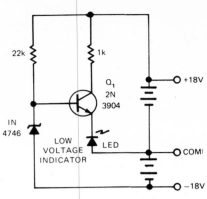

UJT CHARGER FOR 12 V—Keeps 12-V auto storage battery fully charged, for immediate standby use when AC power fails. Power transformer secondary can be 14 to 24 V, rated at about 3 A. Two-gang rotary switch gives choice of three charging rates. Pulse transformer T2 is small audio transformer rewound to have 1:1 turns ratio and about 20 ohms resistance, or can be regular SCR trigger transformer. UJT relaxation oscillator stops when upper voltage limit for battery is reached, as set by pot R6. If oscillator fails to start, reverse one of pulse transformer windings.—F. J. Piraino, Failsafe Super Charger, *73 Magazine*, Holiday issue 1976, p 49.

18-V MONITOR—Circuit turns on LED when ± 18 V battery pack discharges to predetermined low level, while drawing less than 1 mA when LED is off. Zener is reverse-biased for normal operating range of battery. When lower limit is reached, zener loses control and Q_1 becomes forward-biased, turning on LED or other signal device to indicate need for replacement or recharging.—W. Denison and Y. Rich, Battery Monitor Is Efficient, yet Simple, *EDN Magazine*, Oct. 5, 1974, p 76.

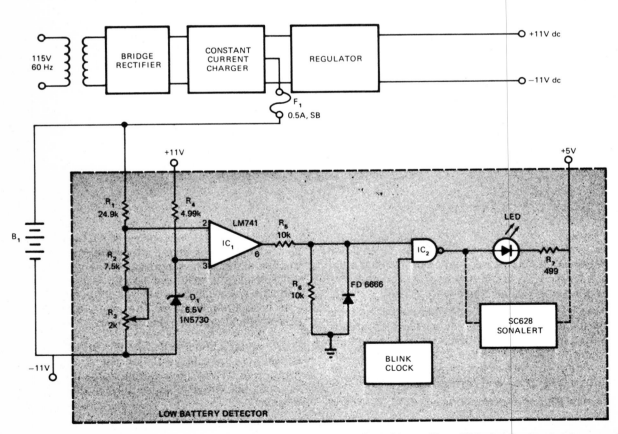

FLASHING LED FOR LOW BATTERY—Developed for use in portable battery-operated test instrument to provide visual indication that depletion level has been reached for series arrangement of 24 nickel-cadmium cells providing 32.5 VDC for regulator of bipolar 11-V supply. Instrument must then be plugged into AC line for recharging of batteries. Voltage across B_1 (nominally 32.5 V) is sensed by R_1-R_4 and D_1. When level drops 24.1 V, opamp comparator output goes positive and enables gate IC_2, so blink clock (such as low-frequency TTL-level oscillator) makes LED flash. Audible alarm is optional.—R. T. Warner, Monitor NiCad's with This Low-Battery Detector, *EDN Magazine*, April 20, 1976, p 112 and 114.

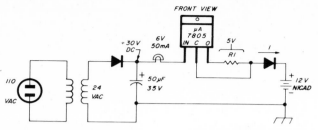

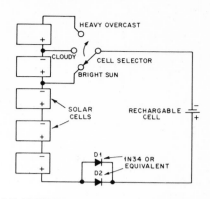

CONSTANT-CURRENT NICAD CHARGER— Constant current is obtained from voltage regulator by floating common line and connecting R1 from output to common terminal. Regulator then tries to furnish fixed voltage across R1. Input voltage must be greater than full-charge battery voltage plus 5 V (for 5-V regulator) plus 2 V (overhead voltage). Changing R1 varies charging current. If R1 is 50 ohms and V is 5 V, constant current is 50 mA through nicad being charged.—G. Hinkle, Constant-Current Battery Charger for Portable Operation, *Ham Radio*, April 1978, p 34—36.

SOLAR-ENERGY CHARGER—Single solar cell on bright day delivers 0.5 V at 50 mA, so three cells are used in bright sun to recharge secondary cell. Switch permits use of additional solar cells on cloudy days. Solar cells can be Radio Shack 276-128.—J. Rice, Charging Batteries with Solar Energy, *QST*, Sept. 1978, p 37.

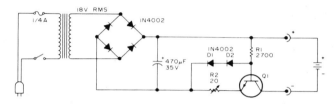

NICAD CHARGER—Regulated charger circuit will handle variable load from 1 to 18 nicad cells. Current-limiting action holds charging current within 1 to 2 mA of optimum value (about one-tenth of rated ampere-hour capacity) from 0 to 24 V. Q1 should have power rating equal to twice supply voltage multiplied by current-limit value. If charging 450-mAh penlight cells, charge current is 45 mA and transistor should be 2 W.—A. G. Evans, Regulated Nicad Charger, *73 Magazine*, June 1977, p 117.

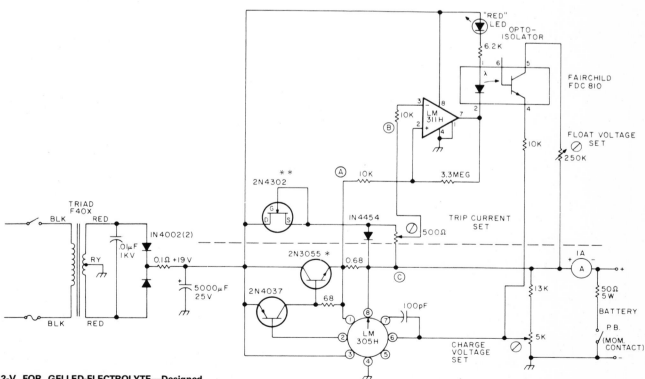

12-V FOR GELLED-ELECTROLYTE—Designed to charge 12-V 3-Ah gelled-electrolyte battery such as Elpower EP1230A at maximum of 0.45 A until battery reaches 14 V, then at constant voltage until charge current drops to 0.04 A. Charger is then automatically switched to float status that maintains 2.2 V per cell or 13.2 V for battery. Circuit is constant-voltage regulator with current limiting as designed around National LM305H, with PNP/NPN transistor pair to increase current capability. Circuit above dashed line is added to standard regulator to meet special charging requirement. Article covers operation and use of circuit in detail.—H. Olson, Battery Chargers Exposed, *73 Magazine*, Nov. 1976, p 98—100 and 102—104.

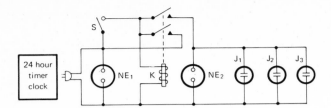

NICAD CHARGE CONTROL—Prevents double-charging if someone forgets to turn off 24-h time clock after recommended 16-h charge period. Nicad devices with built-in chargers are plugged into jacks J_1-J_3, and timer dial is advanced until clock switch is triggered. Neon lamp NE_1 should now come on. Momentary pushbutton switch S is pushed to energize relay K and start charge. When timer goes off, K releases to end charge.—M. Katz, Battery Charge Monitor, *CQ,* July 1976, p 27.

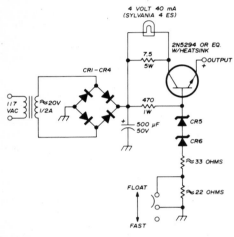

GELLED-ELECTROLYTE BATTERIES—Constant-voltage charger for Globe-Union 12-V gelled-electrolyte storage batteries can provide either fast or float charging. Constant voltage is maintained by series power transistor and series-connected zeners. Output voltage is 13.8 V for float charging and 14.4 V for fast charging.—E. Noll, Storage-Battery QRP Power, *Ham Radio,* Oct. 1974, p 56–61.

CHAPTER 3
Burglar Alarm Circuits

For auto, home, office, and factory installations. Sensors include contact-making, contact-breaking, photoelectric, infrared, Doppler, and sound-actuated devices that trigger circuit immediately or after adjustable delay for driving alarm horn, siren, tone generator, pager, or silent transmitter. Some circuits have automatic shutoff of alarm after fixed operating time as required for auto alarms in some states.

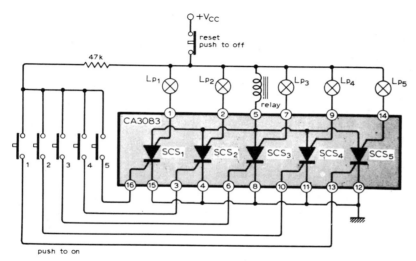

FIVE-INDICATOR ALARM—Single five-transistor IC uses NPN structures on P-type substrate as PNPN silicon controlled switches having common connection for anode (substrate). Relay serving as anode load is energized for actuating alarm if any of the SCS pushbutton switches is closed. Corresponding lamp is energized to identify door or window at which sensor switch has been closed by act of intruder. Alarm remains on until reset by interrupting power supply. Power drain on standby is negligible because SCSs act as open circuits until triggered, permitting use of batteries for supply. Two or more ICs may be added to get more channels.—H. S. Kothari, Alarm System with Position Indication, *Wireless World*, Feb. 1976, p 77.

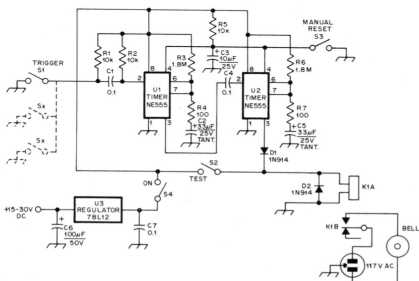

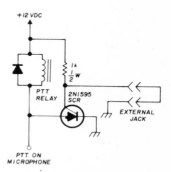

ENTRY-DELAY ALARM—First 555 timer provides delay of about 20 s after triggering by sensor before alarm bell is energized, to allow thief to be caught inside house or give owner time to enter and shut off alarm. Alarm then rings for about 60 s under control of timer U2. Alarm period was set short to attract attention without unduly annoying neighbors.—J. D. Arnold, A Low-Cost Burglar Alarm for Home or Car, *QST*, June 1978, p 35–36.

SCR LATCH—Turns on mobile transceiver or other mobile equipment when power is applied, if external circuit is broken when equipment is stolen. Transmitter will then put unmodulated carrier on air even with PTT switch disconnected or off, for tracing with radio direction finder. If added components are carefully concealed in equipment and new external wiring is worked into existing wiring harness, few thieves will be able to locate trouble. External wires are run under dash so thief must cut them to get out equipment. PTT relay should have protective diode. SCR is 100 PIV, 1 A, but HEP R1003 or R1217 can also be used.—E. Noll, Circuits and Techniques, *Ham Radio*, April 1976, p 40–43.

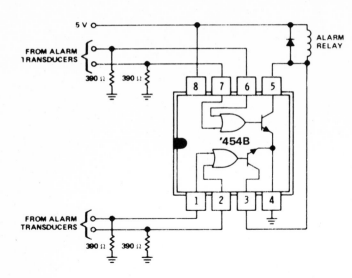

ALARM-SIGNAL DETECTOR—Texas Instruments SN75454B dual peripheral positive-NOR driver energizes alarm relay when alarm signal is received from any one of four different alarm transducers.—"The Linear and Interface Circuits Data Book for Design Engineers," Texas Instruments, Dallas, TX, 1973, p 10-66.

AURAL INDICATOR—Provides attention-getting chirp sound, warble, or continuous tone when turned on by high input from burglar-alarm sensor circuit. Second section of 556 timer provides optional frequency modulation of basic tone to give warbling effect. Chirp is achieved by gating tone oscillator on only during high states of warble oscillator. Aural sensitivity is maximum in range of 1–2 kHz, set by value of R_{t2}.—W. G. Jung, "IC Timer Cookbook," Howard W. Sams, Indianapolis, IN, 1977, p 232–235.

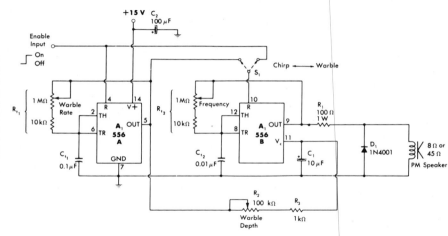

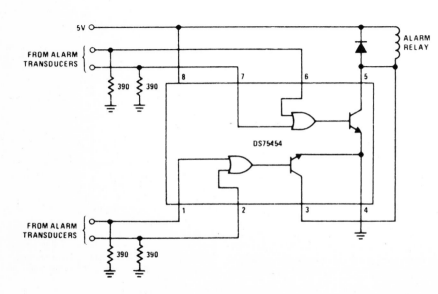

ALARM DETECTOR—National DS75454 dual peripheral NOR driver operating from single 5-V supply energizes alarm relay when one of alarm transducers for either section delivers logic signal as result of intruder action.—"Interface Databook," National Semiconductor, Santa Clara, CA, 1978, p 3-20–3-30.

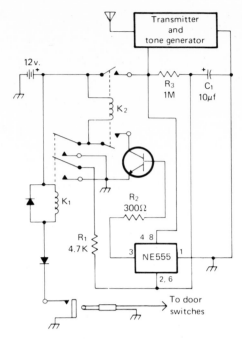

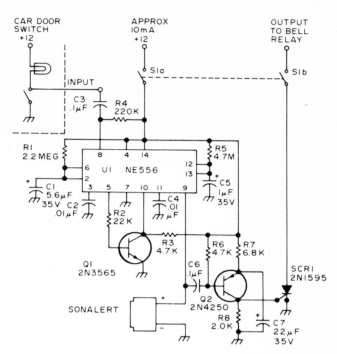

SILENT ALARM—When thief opens car door, relays K_1 and K_2 activate tone-modulated transmitter, which can be any legal combination of power, frequency, and antenna. A few milliwatts of power should be adequate. Thief hears nothing, but owner is alerted via portable receiver tuned to transmitter frequency. Transmitter remains on about 15 s (determined by R_3 and C_1) after door is closed until NE555 times out and removes power from transistor. Use any NPN transistor having adequate current rating for relay. If alarm is provided with its own battery and whip antenna, it cannot be disabled from outside of car.—A. Day, Soundless Mobile Alarm, *CQ,* April 1977, p 11.

CAR-THEFT ALARM—Alarm remains on even if signal from car door switch or other sensor is only momentary, so relay is wired to be self-latching until keyswitch S1 is turned off. Use hood locks or hood-opening sensors to prevent thief from disabling alarm by cutting battery cable. Circuit includes time delay of 6 s for entering car and shutting off alarm, to avoid need for external keyswitch. Sonalert makes loud tone during 6-s delay period to remind driver that alarm needs to be turned off. At end of 6 s, Sonalert stops and much louder bell is energized to further discourage intruder.—J. Pawlan, The Smart Alarm, *73 Magazine,* June 1975, p 37–41.

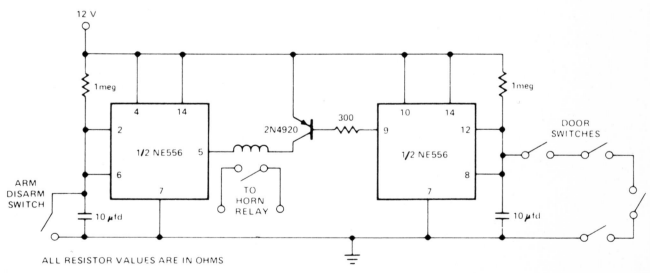

ALL RESISTOR VALUES ARE IN OHMS

SHORT DURATION TIMERS ARE NEEDED TO ALLOW ENTRY AND EXIT

DELAYED ALARM—When normally closed arm/disarm switch is opened, first section of NE556 dual timer starts its timing cycle. After delay to allow for entry or exit, pin 5 goes low to energize alarm circuit. Now, as long as all door switches are closed, PNP transistor is kept off because pin 9 is high. When any door switch is opened, transistor turns on after normal delay for owner to enter car, and horn is sounded unless owner closes arm/disarm switch within delay time.—"Signetics Analog Data Manual," Signetics, Sunnyvale, CA, 1977, p 724–725.

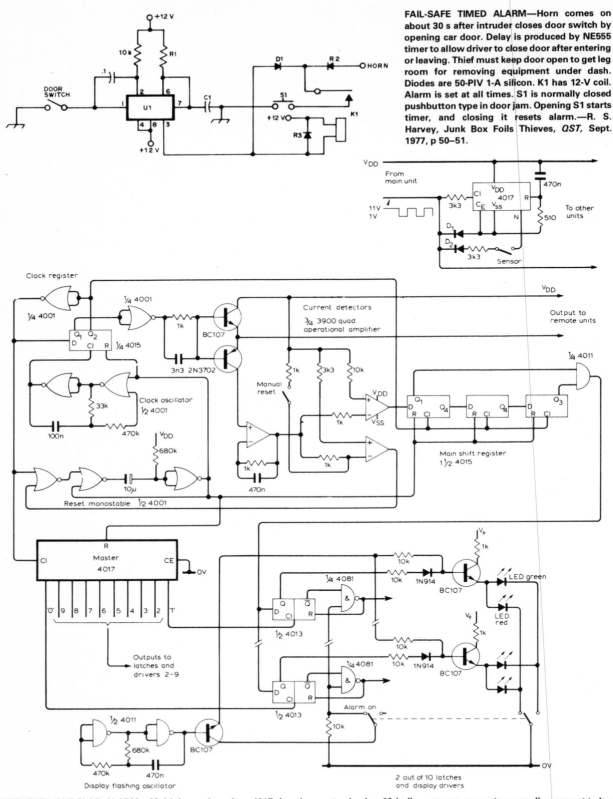

FAIL-SAFE TIMED ALARM—Horn comes on about 30 s after intruder closes door switch by opening car door. Delay is produced by NE555 timer to allow driver to close door after entering or leaving. Thief must keep door open to get leg room for removing equipment under dash. Diodes are 50-PIV 1-A silicon. K1 has 12-V coil. Alarm is set at all times. S1 is normally closed pushbutton type in door jam. Opening S1 starts timer, and closing it resets alarm.—R. S. Harvey, *Junk Box Foils Thieves, QST,* Sept. 1977, p 50–51.

MULTIPLEXED BURGLAR ALARM—Multiplexing technique provides for detection of state of up to 10 sensors, with immediate identification and location of activated sensor. Only one pair of wires runs from control unit to paralleled remote sensor circuits, one of which is shown at upper right. Each sensor location uses different output from one to zero. Multiplexer circuit is based on 4017 decade counter having 10 individual outputs, to give signals in 10 time slots. Power supply rail is used to reset counter. Clock line is eliminated by switching supply line as square wave. Sensor indication line is eliminated by detecting power supply current drain. Control unit uses oscillator and shift register to generate clocking waveforms. 3900 quad opamp converts sensor line current to logic levels for clocking by master 4017 to control 10 output latches and display driver. Two consecutive sensor-open signals are required to activate alarm, minimizing false alarms by interference pulses.—R. J. Chance, *Multiplexed Alarm, Wireless World,* Nov. 1978, p 73–74.

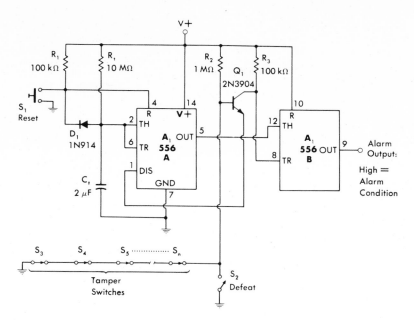

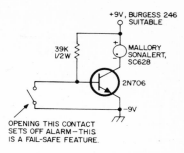

CIRCUIT-BREAKING ALARM—Operates from small 9-V battery, making it independent of AC power failure. Opening of switch or equivalent breaking of foil conductor removes ground from base of transistor, to energize alarm.—Circuits, *73 Magazine,* April 1973, p 132.

WINDOW-FOIL ALARM—Combination of power-up mono MVBR and latch, using both sections of 556 timer, drives output line high when sensor circuit is opened at door or window switch or by breaking foil on glass. Once alarm is triggered, reclosing of sensor has no effect; S_1 must be closed momentarily after restoring sensor circuit to turn alarm off. Circuit includes 22-s power-up delay that prevents triggering of alarm when it is first turned on.—W. G. Jung, "IC Timer Cookbook," Howard W. Sams, Indianapolis, IN, 1977, p 231–232.

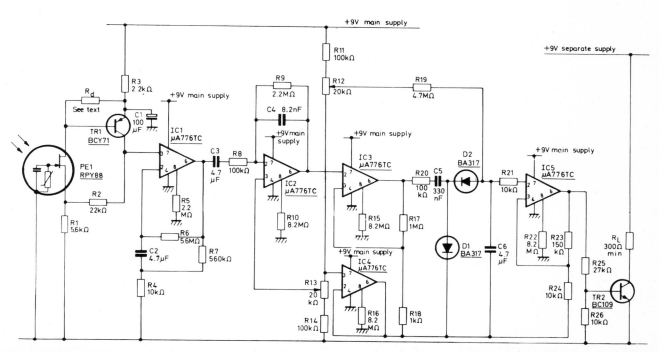

LOW-CURRENT INTRUDER ALARM—Use of programmable μA776 opamps reduces standby current of infrared alarm to 300 μA, permitting operation from small rechargeable cells. Detector is Mullard RPY86 that responds only to wavelengths above 6 μm, making it immune to sunlight and backgrounds intermittently illuminated by sun. Low-cost mirror is used instead of lens to concentrate infrared radiation on detector. R_d is chosen to make input to first opamp between 2 and 6 V. Circuit energizes alarm relay R_L only when incident radiation is changed by movement of intruder in monitored space.— "Ceramic Pyroelectric Infrared Detectors," Mullard, London, 1978, Technical Note 79, TP1664, p 8.

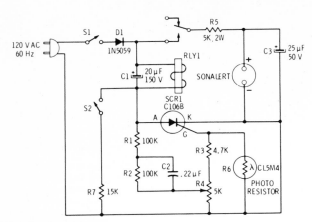

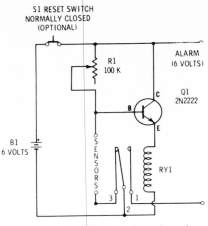

LIGHT-INTERRUPTION DETECTOR—Use of SCR as regenerative amplifier rather than as switch gives extremely high sensitivity to very slight reductions in light reaching photoresistor. Requires no light source or accurately aligned light-beam optics. In typical application as burglar alarm, light shining through window from streetlight provides sufficient ambient illumination so any movement of intruder within 10 feet of unit will energize Sonalert alarm. Sensitivity control R4 is adjusted so SCR receives positive pulses from AC line, but their ampli-

tude is not quite enough to start regenerative action of SCR. Reduction in light then increases resistance of photoresistor enough to raise level of gate pulses for SCR, starting regenerative amplification that energizes relay. Use Mallory SC-628P Sonalert which produces pulsed 2500-Hz sound. With S2 open, alarm stops when changes in light cease. With S2 closed, alarm is latched on and S1 must be opened to stop sound.—R. F. Graf and G. J. Whalen, "The Build-It Book of Safety Electronics," Howard W. Sams, Indianapolis, IN, 1976, p 7–12.

LATCHING ALARM—Closed-circuit alarm drawing only 130 μA of standby current from battery is turned on by opening sensor switch or cutting wire. Automatic latching contacts on relay prevent burglar or intruder from deactivating alarm by resetting sensor switch. Relay is Radio Shack 275-004. Sensor can be foil strip around window subject to breakage.—F. M. Mims, "Transistor Projects, Vol. 3," Radio Shack, Fort Worth, TX, 1975, p 75–86.

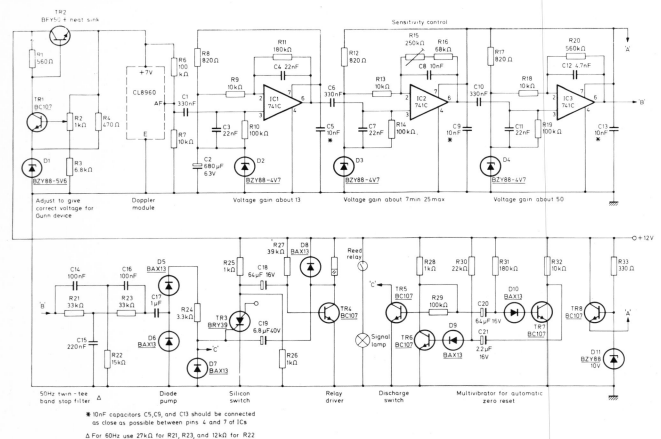

* 10nF capacitors C5, C9, and C13 should be connected as close as possible between pins 4 and 7 of ICs

Δ For 60Hz use 27kΩ for R21, R23, and 12kΩ for R22

MICROWAVE DOPPLER INTRUSION ALARM—Mullard CL8960 X-band Doppler radar module detects movement of remote target by monitoring Doppler shift in microwave radiation reflected from target. Module consists of Gunn oscillator cavity producing energy to be radiated, mounted alongside mixer cavity that

combines reflected energy with sample of oscillator signal. Transmitted frequency is 10.7 GHz. Doppler change is about 31 Hz for relative velocity of 0.45 m/s (1 mph) of relative velocity between object and module, giving AF output for velocities up to 400 mph. Filtered AF is applied through diode pump to trigger of silicon

controlled switch TR3 that makes contacts of reed relay open for about 1 s. Relay action is repeated as long as intruder is in monitored area. Report covers circuit operation in detail.—J. E. Saw, "Microwave Doppler Intruder Alarms," Mullard, London, 1976, Technical Information 36, TP1570, p 6.

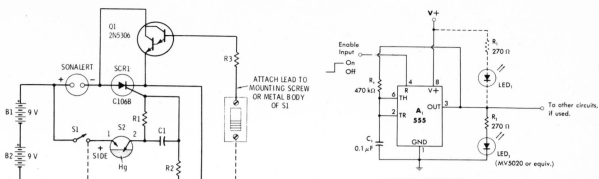

VISUAL INDICATOR—When circuit is activated by high output of burglar alarm circuit, 555 timer operating as very low frequency MVBR makes LED$_1$ flash on and off during alarm condition. Alternate connection of LED$_1$ to V+ holds LED$_1$ on for standby while flashing it during alarm. Oscillator output is also available for other uses if desired. Indicator can be located remotely from alarm.—W. G. Jung, "IC Timer Cookbook," Howard W. Sams, Indianapolis, IN, 1977, p 232–235.

NOTE: TOUCHING MOUNTING SCREW OF S1 AND CASE OF S2 COMMUTATES SCR BY MOMENTARILY ENERGIZING Q1

HOTEL-ROOM ALARM—Alarm mounted in flashlight-shaped cylinder is positioned on floor inside hotel room in such a way that it is knocked over by intruder opening door. Mercury switch S2 then triggers SCR and activates Mallory SC-628P pulsed Sonalert alarm. Circuit latches on and can be turned off only by use of Darlington-amplifier touch switch. Connection from base of Darlington to positive terminal of battery must be made through fingertips as shown by dashed line in order to silence alarm. Once silenced, S1 can be opened to disconnect latch so alarm can be moved. Other applications include protection of unattended luggage. C1 is 0.1 µF, R1 is 1 megohm, R2 is 1K, R3 is 39K, and S2 is mercury element removed from GE mercury toggle switch.—R. F. Graf and G. J. Whalen, "The Build-It Book of Safety Electronics," Howard W. Sams, Indianapolis, IN, 1976, p 19–24.

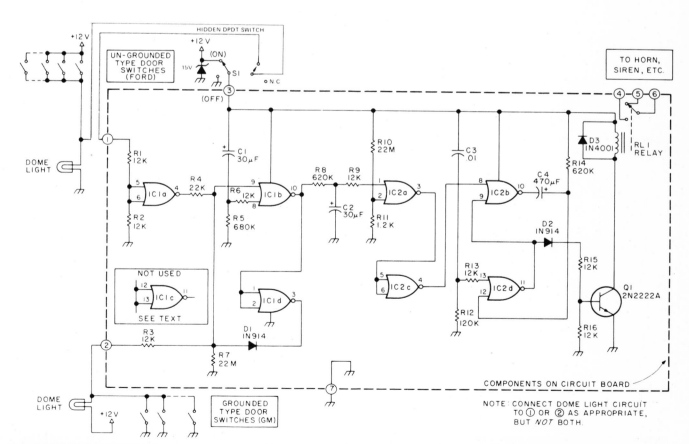

5-min SHUTOFF—Vehicle intrusion alarm shuts off automatically in about 5 min after being triggered, as required by law in some states. Drain on battery is negligible until alarm is set off by intruder. Once triggered, operation sequence is not affected by subsequent opening or closing of doors. System uses two CMOS CD4001AE quad two-input NOR gates for switching logic. IC1 provides sensor interface, latch, and entry/exit time delays. IC2 provides output through Q1 and relay, as well as automatic shutoff delay. Article gives construction details and layout for printed-circuit board.—W. J. Prudhomme, Vehicle Security Systems, *73 Magazine*, Oct. 1977, p 122–125.

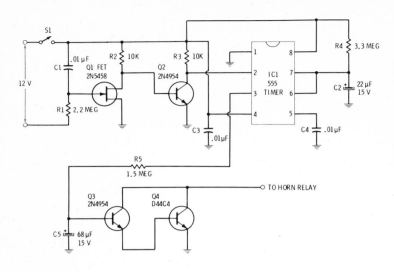

CURRENT-DRAIN SENSOR—Current drawn by dome light when door is open or by ignition when turned on triggers current-sensing stages Q1 and Q2 to start 555 timer and apply power to horn relay. Initial 15-s delay in sounding horn allows owner to enter car and open hidden switch S1 to deactivate alarm. If S1 is not opened during delay interval, horn sounds for about 90 s, then circuit automatically resets itself. C5 and R5 control duration of initial 15-s delay. C2 and R4 control total time that horn sounds.—R. F. Graf and G. J. Whalen, "The Build-It Book of Safety Electronics," Howard W. Sams, Indianapolis, IN, 1976, p 57–62.

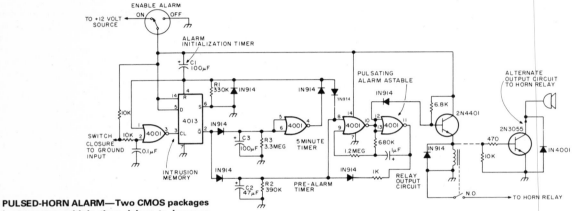

PULSED-HORN ALARM—Two CMOS packages incorporate multiple time delays to improve convenience and effectiveness of auto intrusion alarm. R1C1 gives 30-s delay for arming alarm after it is turned on by switch concealed inside car, to let driver get out of car. R2C2 gives 15-s delay before alarm sounds after door is opened, to allow driver to get back in car again and disable alarm. R3C3 turns off alarm in 300 s and resets alarm system for next intrusion. Car horn is pulsed 60 times per minute, so alarm would not be confused with stuck horn. Article tells how circuit works and gives detailed instructions for installation and connection to door and trunk switches.—G. Hinkle, Give the Hamburglar Heart Failure, *73 Magazine*, Feb. 1977, p 36–37.

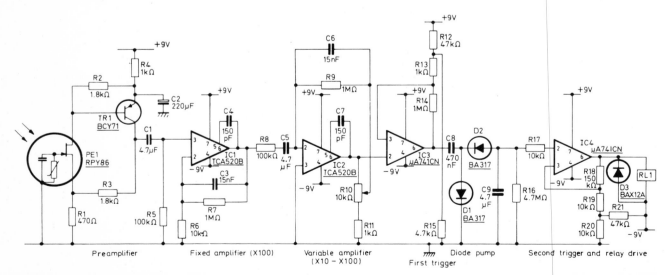

INTRUDER ALARM—Input is from Mullard RPY86 infrared detector responding to wavelengths above 6 μm, making it immune to sunlight and backgrounds intermittently illuminated by sun. Output signal is produced only when incident radiation is changed by movement of intruder in monitored space. Mirrors rather than lenses concentrate incident radiation on detector because mirrors do not require high-quality surface finish. Preamp is followed by two amplifier stages, with R10 varying gain of second stage between 10 and 100. Bandwidth is 0.3–10 Hz. First trigger, having threshold of about 1 V, drives second trigger through diode pump to energize alarm relay when intruder is present.—"Ceramic Pyroelectric Infrared Detectors," Mullard, London, 1978, Technical Note 79, TP1664, p 8.

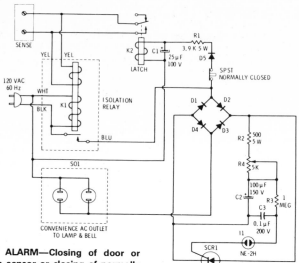

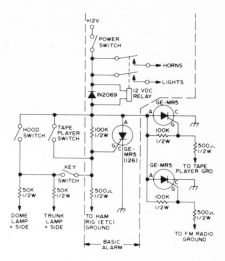

OPEN-CIRCUIT ALARM—Closing of door or window switch sensor or closing of normally open panic-button switch at bedside and other strategic locations in home trips alarm that sounds loud bell and flashes bright light on and off. Sensor shorts control winding of K1, allowing K1 to drop out and apply line voltage to alarm circuit. One AC path is through D5 which rectifies AC for energizing DC latch relay K2 to short sensor lines even though initiating sensor has opened. Simultaneously, AC is applied to diode bridge having SCR between DC legs. C2 starts charging through R2 and R4, and C3 charges through R3. When voltage across C3 reaches about 90 VDC, it fires neon and C3 discharges into gate of SCR. Full line voltage is then applied to lamp and bell plugged into load outlets. When C2 drops below holding current, SCR turns off during next AC cycle and load goes off until neon fires again. Setting of 5K pot R4 gives range of 15-80 flashes and horn pulses per second. To stop alarm, open SPST switch momentarily.—R. F. Graf and G. J. Whalen, "The Build-It Book of Safety Electronics," Howard W. Sams, Indianapolis, IN, 1976, p 75–80.

WIRE-CUTTING ALARM—SCR normally acts as open circuit in series with 12-VDC alarm relay because grid is made negative by voltage divider consisting of 100K in series with 500 ohms. If ground on 500-ohm resistor is removed, as by removal of tape player or CB set from car by thief, gate becomes more positive and SCR conducts, to energize relay, sound horn, and make headlights shine brightly. Additional triggering SCRs or alarm switches can be added as shown outside of dashed area for basic alarm.—A. Szablak, Another Burglar Alarm, *73 Magazine*, May 1974, p 45–46.

SOUND-ACTIVATED SWITCH—Can be used as sensor for burglar alarm or for turning on surveillance tape recorder to monitor conversations. R_8 is adjusted to give desired sensitivity at which A_2 triggers switch Q_1 to provide 200-mA load current and turn on indicator LED. First section of LM339 quad comparator serves as amplifier and detector providing gain of 100. Second comparator compares DC output of first with reference level selected by R_8.—D. R. Morgan, Sound Turns Switch On, *EDN Magazine*, Aug. 5, 1978, p 82 and 84.

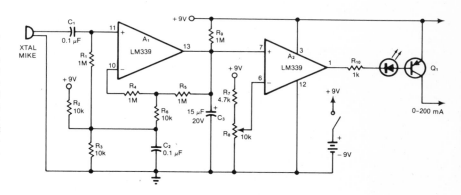

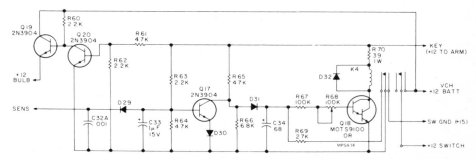

ALARM DRIVES PAGING BEEPER—Complete protection of vehicle is provided by multiplicity of door-switch, mat-switch, vibration, motion, and other sensors connected to common sensor input of alarm switching circuit that controls radio pager, 1-W GE Voice Command II trans- mitter operating around 147 MHz, 100-W electronic siren, and power horns. Closing of contacts in any sensor grounds common input (assuming keylock switch has been closed to arm circuit by applying +12 V), applying power to siren and pager system. Range is about 1 mi for Motorola Pageboy II cigarette-pack-size pager receiver. Article describes construction, operation, and installation in detail and gives complete circuit of pager.—J. Crawford, Build a Beeper Alarm, *73 Magazine*, Oct. 1977, p 68–77.

DOPPLER BURGLAR ALARM—Small radar transmitter operating at 10.687 GHz fills protected area with radio waves. Waves reflected from stationary objects are ignored by receiver, while waves undergoing Doppler shift in frequency by reflection from moving object such as intruder are selectively amplified for triggering of alarm. Single waveguide section is divided into two cavities, each having Gunn diode; transmitter cavity feeds points A and B of transmitter TR7-IC₃, and other cavity feeds points C and D of amplifier that drives alarm relay. Article covers construction and operation of circuit and gives sources (British) for parts and construction kits. Opamps are SN72748 or equivalent, IC₃ is μA723 or equivalent, Tr₁-Tr₃ are ZTX500 or equivalent, Tr₄-Tr₆ are ZTX302 or equivalent, Tr₇ is 3055, D₁-D₈ are 1N4001 or equivalent, D₉-D₁₀ are 1N914, SCR₁ is TIC44 or equivalent, Z₁-Z₂ are BZY88-C8V2, relay is 18-V with 1K coil, Doppler module is Mullard CL8960 or equivalent, and self-oscillating mixer for receiver is Mullard CL8630S or equivalent. Alarm stays on until reset by appropriate switch.—M. W. Hosking, Microwave Intruder Alarm, *Wireless World,* July 1977, p 36–39.

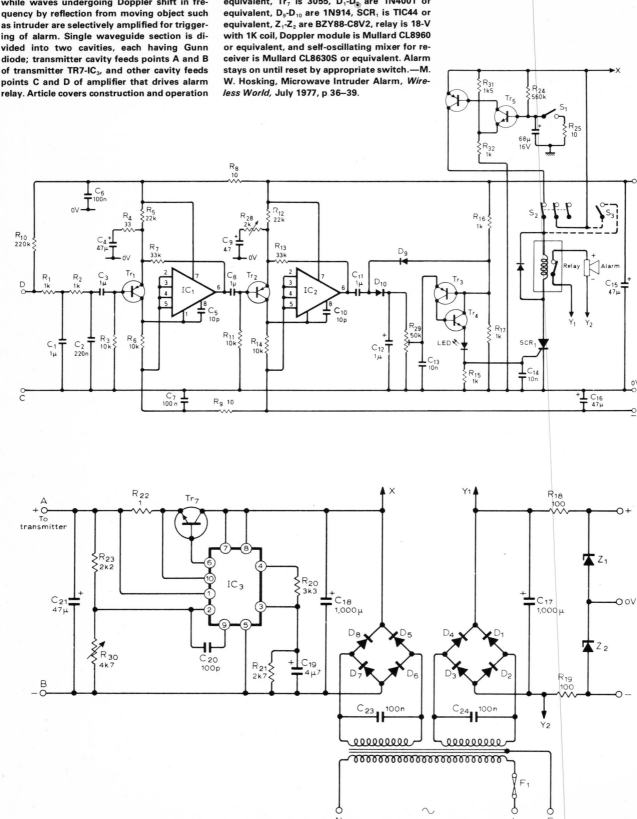

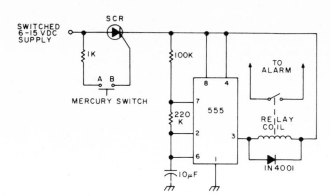

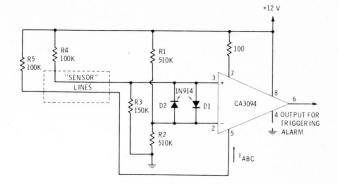

BEEPER—Intermittent alarm using 555 timer can be set to energize horn, lights, or other signaling device at any desired interval when tripped. When used on auto, sound cannot be mistaken for stuck horn. Choose SCR rating to handle current drawn by relay and timer. If alarm draws less than 200 mA, relay is not needed.—W. Pinner, Alarm! Alarm! Alarm!, *73 Magazine,* Feb. 1976, p 138–139.

OPEN/SHORT/GROUND ALARM—Pin 6 of CA3094 IC is high for no-alarm condition. When any one sensor line is open, is shorted to other line, or is shorted to ground, output of IC goes low and resulting output current serves for activating alarm system.—E. M. Noll, "Linear IC Principles, Experiments, and Projects," Howard W. Sams, Indianapolis, IN, 1974, p 316–317.

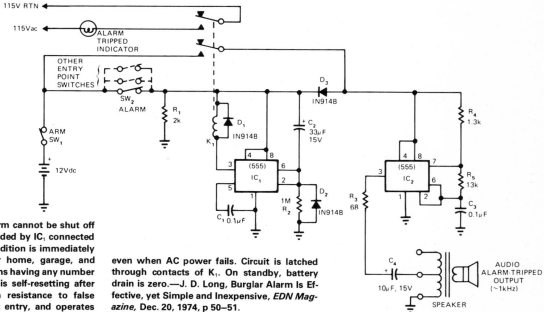

LATCH-ON ALARM—Alarm cannot be shut off for 12 s, with delay provided by IC₁ connected as mono, even if trip condition is immediately removed. Developed for home, garage, and auto burglar alarm systems having any number of trip switches. Circuit is self-resetting after delay interval, has high resistance to false alarms other than direct entry, and operates even when AC power fails. Circuit is latched through contacts of K₁. On standby, battery drain is zero.—J. D. Long, Burglar Alarm Is Effective, yet Simple and Inexpensive, *EDN Magazine,* Dec. 20, 1974, p 50–51.

CHAPTER 4
Digital Clock Circuits

Provide 12- or 24-hour time on LED, LCD, gas-discharge, or fluorescent digital displays for watches and clocks. Some also have calendar display and alarm-tone generator. Special circuits provide battery backup for AC power failure, multiplexing of display to reduce battery drain, stopwatch, and tide clock.

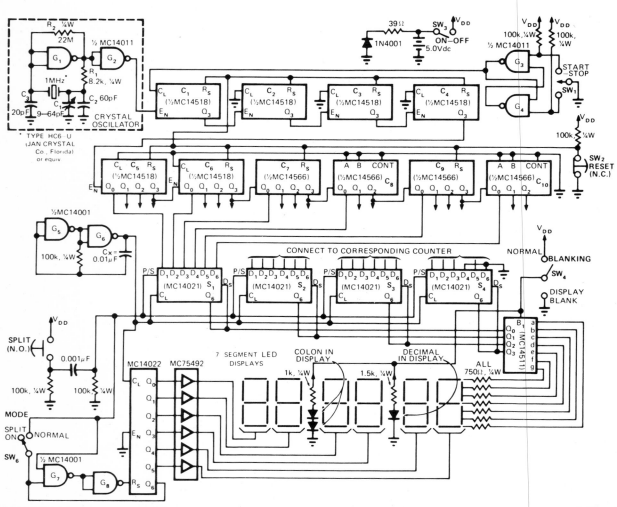

6-DIGIT STOPWATCH—Low-cost battery-powered electronic stopwatch with 6-digit LED display uses readily available complex-function CMOS ICs to minimize component count. Time range is up to 59 min and 59.99 s. Multiplexing by time-sharing counters through one display-driving decoder cuts battery drain because each digit is on for only one-sixth of time. Article traces operation of circuit step by step. Maximum error is only 0.001 s/h. Four rechargeable nicad batteries last 500 h per charge if displays are blanked when not being read, and about 6 h without blanking.—A. Mouton, Build Your Own Digital Stopwatch with Strobed LED Read-out, *EDN Magazine*, April 5, 1974, p 55–57.

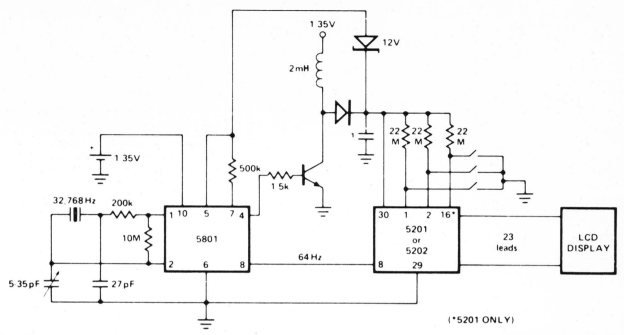

(*5201 ONLY)

LCD WRISTWATCH—Inverter section of Intel 5801 oscillator/divider is used with 32,768-Hz crystal to produce time base. First divider in 5801 reduces this to 1024 Hz for driving upconverter transistor. Feedback from transistor through 12-V zener is used to regulate and control pulse width of 1024-Hz signal. Upconverter also provides 12–15 V required by LCD and 5201 decoder/driver IC. Output to each LCD segment and to common backplate is 32-Hz square wave. Separate drive flashes colon at 1-Hz rate.—M. S. Robbins, "Electronic Clocks and Watches," Howard W. Sams, Indianapolis, IN, 1975, p 128–130.

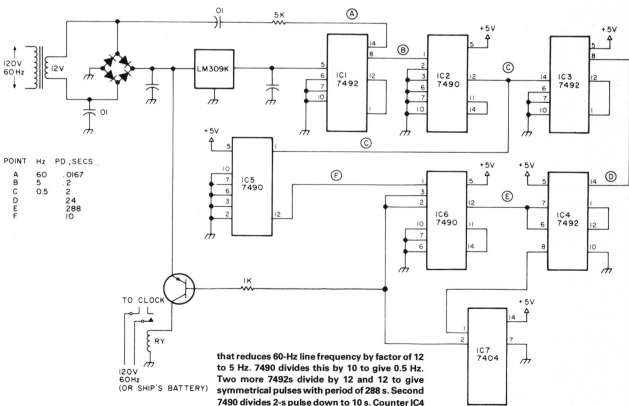

POINT	Hz	PD.,SECS
A	60	.0167
B	5	.2
C	0.5	2
D		24
E		288
F		10

TIDE CLOCK—Circuit shuts off electric clock of any type for 5 s out of every 144 s, to give loss of 50 min in 24 h as required for making high tides conform to clock readings. Regulated 5-V supply shown drives TTL 7492 frequency divider that reduces 60-Hz line frequency by factor of 12 to 5 Hz. 7490 divides this by 10 to give 0.5 Hz. Two more 7492s divide by 12 and 12 to give symmetrical pulses with period of 288 s. Second 7490 divides 2-s pulse down to 10 s. Counter IC4 inhibits 5-s counter by feeding low output into one gate of IC7 hex inverter. When IC4 counts up to 144 s, its output goes high and resets IC6 to low for start of 5-s low period of that counter. Article gives timing waveforms. Switching transistor is used to control relay that opens clock circuit. Set tide clock at 12:00 for high tide at location of use, and it will be 12:00 at high tide thereafter. Low tide will then be at 6:00.—J. F. Crowther, Time and Tide—Digitally, *73 Magazine,* Aug. 1978, p 156–157.

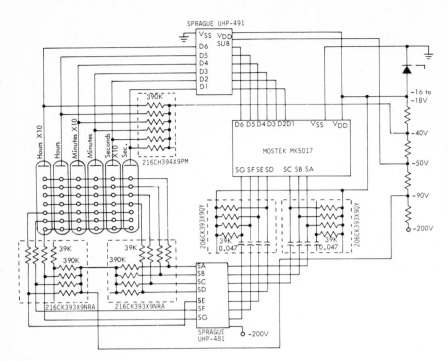

12-h WITH SECONDS—Combination of Mostek clock IC and Sprague high-voltage display drivers, acting through 206C and 216C single in-line resistor network, provides drive for conventional seven-element gas-discharge digital clock display showing hours, minutes, and seconds. Requires −200 V supply. Display can be Burroughs Panaplex, Cherry Plasma-Lux, or Beckman SP series.—"Integrated Circuits Data Book—1," Sprague, North Adams, MA, 1978, p 3-5.

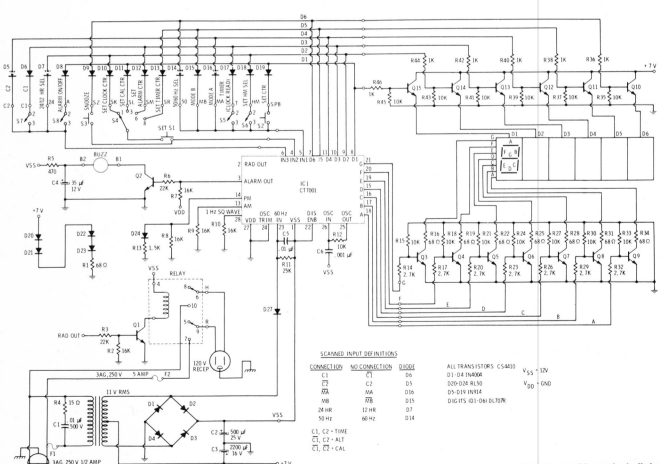

6-DIGIT WITH CALENDAR AND ALARMS—Circuit is built around Cal-Tex CT7001 IC that includes outputs for displaying day of month along with time on Litronix DL707 LED readouts. Transistor switch Q1 and relay form timer triggered by IC to control radio or other appliance drawing up to 5 A from AC line. Dual-voltage power supply provides 7 and 14 VDC. Includes snooze alarm along with regular built-in transistor-driven buzzer.—M. S. Robbins, "Electronic Clocks and Watches," Howard W. Sams, Indianapolis, IN, 1975, p 103–104 and 116–117.

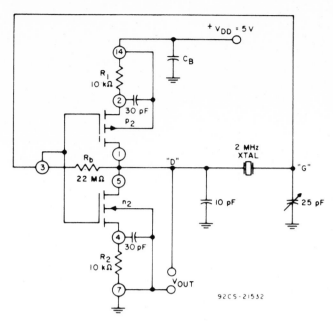

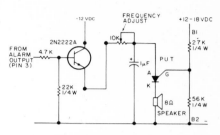

2-MHz CRYSTAL USING CMOS PAIR—One CMOS transistor pair from CA3600E array is connected with feedback pi network to give stable oscillator performance with 2-MHz crystal.

Low power drain makes circuit ideal for use in digital clocks and watches.—"Linear Integrated Circuits and MOS/FET's," RCA Solid State Division, Somerville, NJ, 1977, p 280.

ALARM FOR DIGITAL CLOCK—Uses transistor as driver to turn on programmable unijunction transistor (PUT) oscillator feeding 8-ohm loudspeaker. Pitch of tone can be adjusted with 10K pot. Input is from alarm pin of digital clock IC (pin 3 for Fairchild FCM7001 equivalent of Cal-Tex CT7001). PUT is Radio Shack 276-119 or equivalent.—W. J. Prudhomme, CT7001 Clockbuster, *73 Magazine*, Dec. 1976, p 52–54 and 56–58.

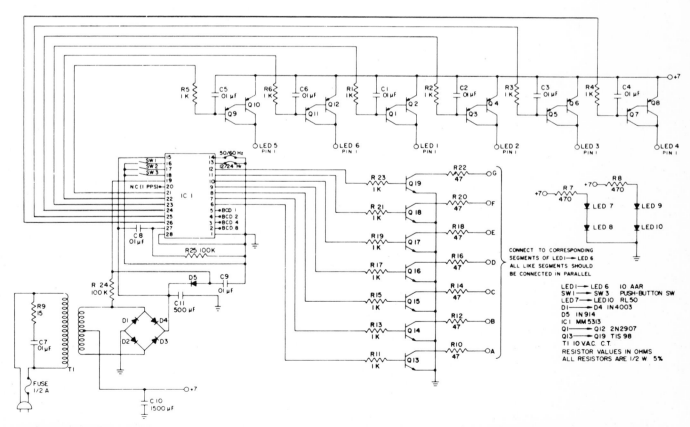

6-DIGIT LED WITH SLEW BUTTONS—National MM5313 PMOS digital clock IC drives display which includes four discrete LEDs mounted on readout panel to form colons between hours, minutes, and seconds. AC supply provides 14 VDC for IC and 7 VDC for displays. Hold push-button SW1 stops count to give precise seconds setting. Slow-slew button SW2 advances time at 1 min/s for precise setting, and fast-slew button SW3 advances time 1 h/s. Digit drivers Q1-Q12 are Darlington-connected pairs of PNP transistors. Segment drivers Q13-Q19 are single PNP transistors.—M. S. Robbins, "Electronic Clocks and Watches," Howard W. Sams, Indianapolis, IN, 1975, p 103 and 113.

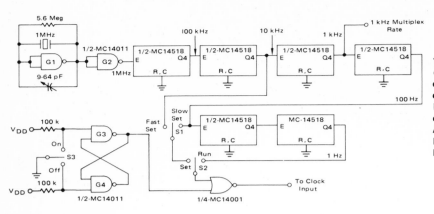

1-Hz REFERENCE—Output of 1-MHz crystal oscillator is stepped down to 1 Hz by CMOS decade divider chain using Motorola MC14518 dual decade counters. Circuit also generates 1-kHz multiplex rate for display used with 24-h industrial clock. Supply is +5 V.—D. Aldridge and A. Mouton, "Industrial Clock/Timer Featuring Back-Up Power Supply Operation," Motorola, Phoenix, AZ, 1974, AN-718A, p 5.

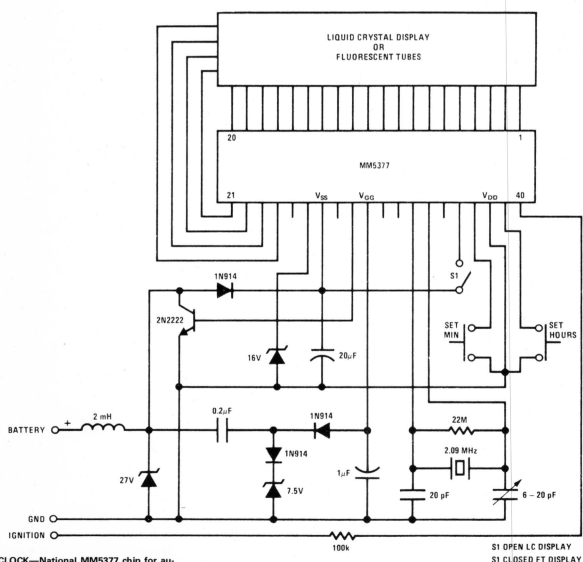

AUTO CLOCK—National MM5377 chip for automobile clock interfaces directly with 4-digit liquid-crystal or fluorescent-tube display. 12-h format includes leading-zero blanking and colon indication. Voltage-sensitive output drives energy-storage network serving as voltage doubler/regulator. Crystal oscillator is referenced time base.—"MOS/LSI Databook," National Semiconductor, Santa Clara, CA, 1977, p 1-33–1-37.

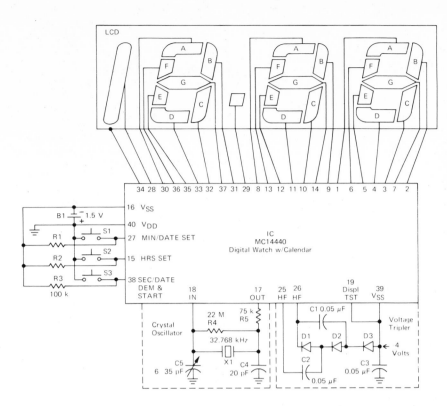

1.5-V LCD—Will operate over 1 year on single 1.5-V AAA battery with accuracy of ±1 min. Basic timekeeping functions are provided by Motorola MC14440 CMOS device that includes calendar. 32.768-kHz NT-cut quartz crystal and trimming capacitor provide reference frequency. Output of 1.5-V alkaline cell is increased to 4 V for display by voltage tripler using MBD101 Schottky diodes.—J. Roy and A. Mouton, "A Cordless, CMOS, Liquid-Crystal Display Clock," Motorola, Phoenix, AZ, 1977, EB-56.

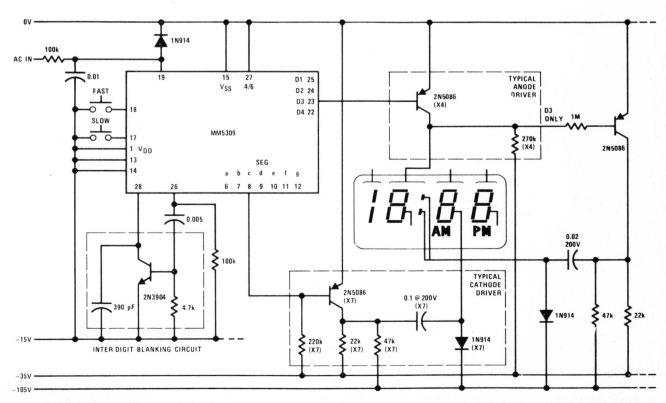

GAS-DISCHARGE DISPLAY—National MM5309 digital clock gives choice of 12- or 24-h display and 50- or 60-Hz operation for driving 4-digit gas-discharge display having colon and AM/PM indications. Separate cathode driver and separate anode driver are required for each digit.— "MOS/LSI Databook," National Semiconductor, Santa Clara, CA, 1977, p 1-2–1-8.

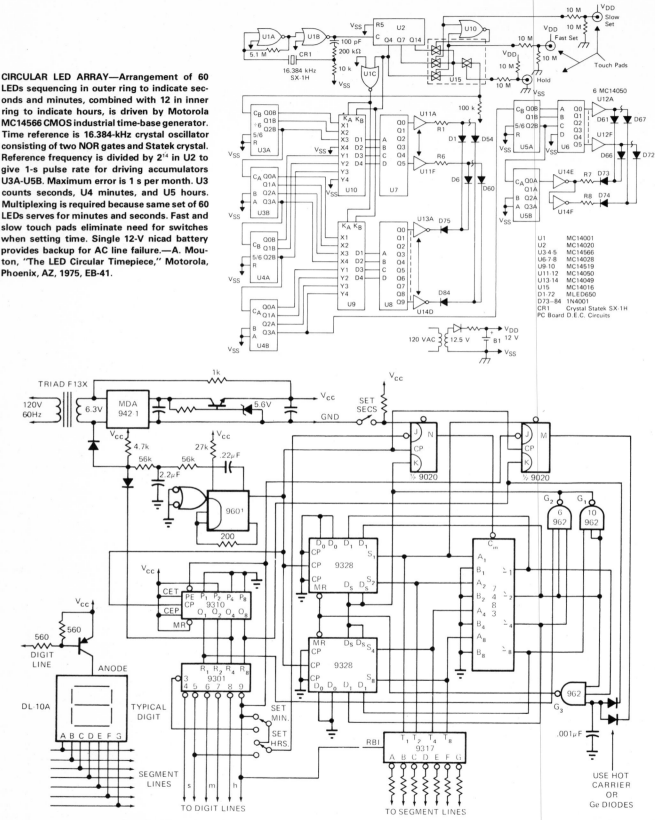

CIRCULAR LED ARRAY—Arrangement of 60 LEDs sequencing in outer ring to indicate seconds and minutes, combined with 12 in inner ring to indicate hours, is driven by Motorola MC14566 CMOS industrial time-base generator. Time reference is 16.384-kHz crystal oscillator consisting of two NOR gates and Statek crystal. Reference frequency is divided by 2^{14} in U2 to give 1-s pulse rate for driving accumulators U3A-U5B. Maximum error is 1 s per month. U3 counts seconds, U4 minutes, and U5 hours. Multiplexing is required because same set of 60 LEDs serves for minutes and seconds. Fast and slow touch pads eliminate need for switches when setting time. Single 12-V nicad battery provides backup for AC line failure.—A. Mouton, "The LED Circular Timepiece," Motorola, Phoenix, AZ, 1975, EB-41.

MULTIPLEXED CLOCK DISPLAY—Multiplexed display suitable for LED readouts is provided by circuit using TTL counters to count 60-Hz line. When count reaches 10 o'clock, flip-flop M is set on every cycle. Gate G_3 then detects when time goes to 13 o'clock, and clears shift register. Carry flip-flop remains set, so 1 is loaded into hours digit to accomplish transition from 12:59:59 to 1:00:00. Seven-segment decoder driver looks at shift register output and drives segment lines of LED. Leading hours digit is blanked, using RBI input on 9317.—G. Smith, Novel Clock Circuit Provides Multiplexed Display, *EDN Magazine,* Sept. 1, 1972, p 50–51.

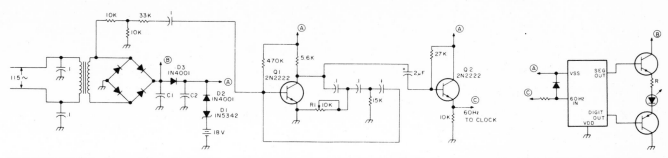

STANDBY SUPPLY—Phase-shift oscillator Q1 operates from AC line through bridge-rectifier power supply and provides line-synchronized 60-Hz power to standard digital clock through isolating emitter-follower Q2. During power outage, oscillator is switched automatically to battery by diode network and provides reasonably accurate signal for operating clock. Free-running oscillator is adjusted to be slightly low, such as 59.9 Hz. For reasonably long power outage, say 4 h, this 0.1-Hz error is equivalent to 0.167% error in time, so clock loses only 24 s during outage. C1 and C2 are 200 to 300 μF. Adjust R1 to give output just below 60 Hz on battery operation. To minimize battery drain, LEDs on digital clock are not energized during standby.—R. S. Isenson, Digital Clock Fail-Safe, *73 Magazine*, July 1977, p 168–169.

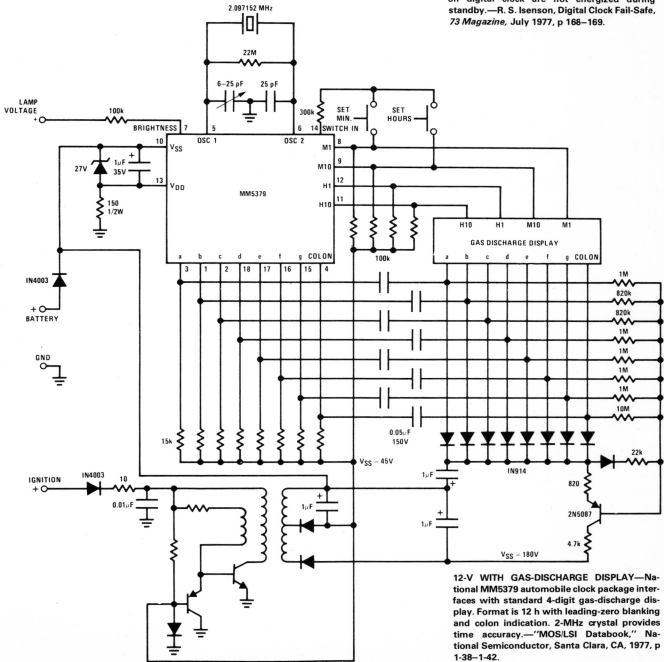

12-V WITH GAS-DISCHARGE DISPLAY—National MM5379 automobile clock package interfaces with standard 4-digit gas-discharge display. Format is 12 h with leading-zero blanking and colon indication. 2-MHz crystal provides time accuracy.—"MOS/LSI Databook," National Semiconductor, Santa Clara, CA, 1977, p 1-38–1-42.

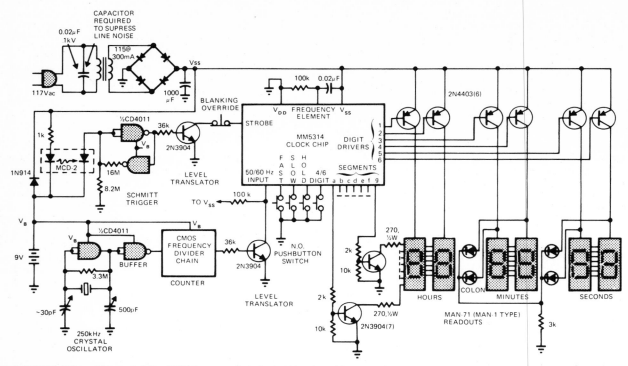

AC/DC CLOCK—When AC power fails, MCD-2 optoisolator senses voltage drop and makes Schmitt trigger force strobe input of clock chip to ground, blanking display and reducing cur- rent drain from 200 mA on AC to 12 mA on 9-V standby battery. Clock will run for days on 1000- mAh battery. Two LED pairs that form colons between time digits are operated from digit strobe lines and remain lit when display is blanked, but draw only 1 mA.—S. I. Green, Dig- ital Clock Keeps Counting Even When AC Power Fails, *EDN Magazine,* Dec. 20, 1974, p 49–51.

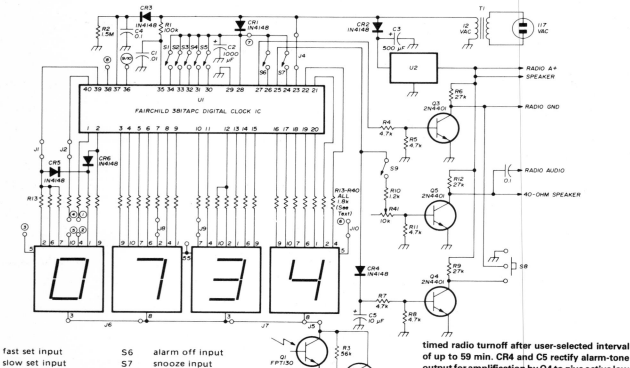

S1	fast set input	S6	alarm off input
S2	slow set input	S7	snooze input
S3	seconds display input	S8	alarm tone on/off
S4	alarm display input	S9	alarm output on/off
S5	sleep display input	R41	tone amplitude control

DIGITAL ALARM—Direct drive offered by Fair- child 3817 IC allows design of simple low-cost clock radio providing display drive, alarm, and sleep-to-music features in 12- or 24-h formats. Display is Fairchild FND500 LED. Either 50- or 60- Hz input may be used. U2 is 7800-series IC volt- age regulator rated to meet requirements of radio used. Q3 provides active low output for timed radio turnoff after user-selected interval of up to 59 min. CR4 and C5 rectify alarm-tone output for amplification by Q4 to give active low output for timed radio turn-on when coinci- dence is detected by alarm comparators. Q5 provides alarm-tone output at level sufficient to drive 40-ohm loudspeaker with ample wake-up volume. If radio is used, omit loudspeaker. Ar- ticle covers construction and adjustment.—D. R. Schmieskors, Jr., Low-Cost Digital Clock, *Ham Radio,* Feb. 1976, p 26–30.

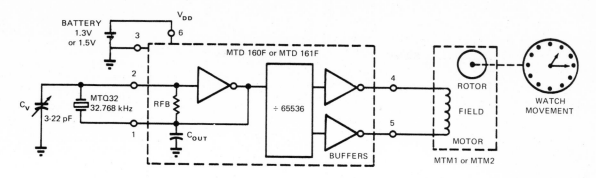

C_V = Trimmer capacitance
C_{OUT} = Integrated oscillator output capacitance
$\approx$ 20 pF
R_{FB} = Integrated oscillator feedback resistance
$\approx$ 40 M

QUARTZ-MOTOR WRISTWATCH—Uses one 32.768-kHz crystal at input of Motorola MTD 160F or 161F custom CMOS chip, with stepper motor at output of chip for driving conventional watch hands. Chip contains three-inverter oscillator, 16 counting flip-flops, and motor drive buffers.—B. Furlow, CMOS Gates in Linear Applications: The Results Are Surprisingly Good, *EDN Magazine,* March 5, 1973, p 42—48.

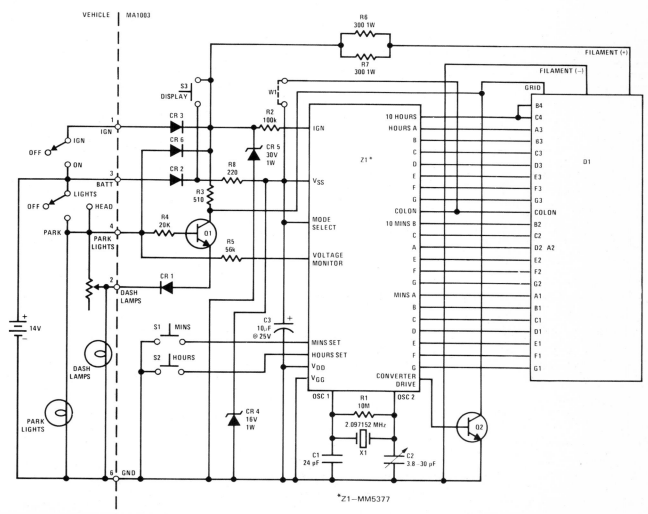

12-V AUTO CLOCK—National MA1003 automotive/instrument clock module combines MM5377 MOS LSI clock with 4-digit 0.3-inch green vacuum fluorescent display, 2.097-MHz crystal, and discrete components on single printed-circuit board to give complete digital clock. Brightness control logic blanks display when ignition is off, reduces brightness to 33% when parking or headlight lamps are on, and follows dash-lamp dimming control setting. Display has leading-zero blanking. For portable applications, display can be activated by closing display switch momentarily.—"MOS/LSI Databook," National Semiconductor, Santa Clara, CA, 1977, p 13-8—13-10.

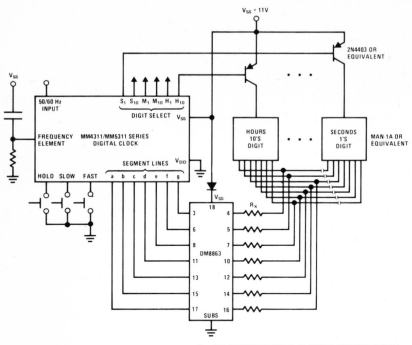

6-DIGIT DISPLAY—National DM8863 8-digit LED driver serves as segment driver for common-anode display of hours, minutes, and seconds, replacing total of 14 resistors and 7 transistors.—C. Carinalli, "Driving 7-Segment LED Displays with National Semiconductor Circuits," National Semiconductor, Santa Clara, CA, 1974, AN-99, p 11.

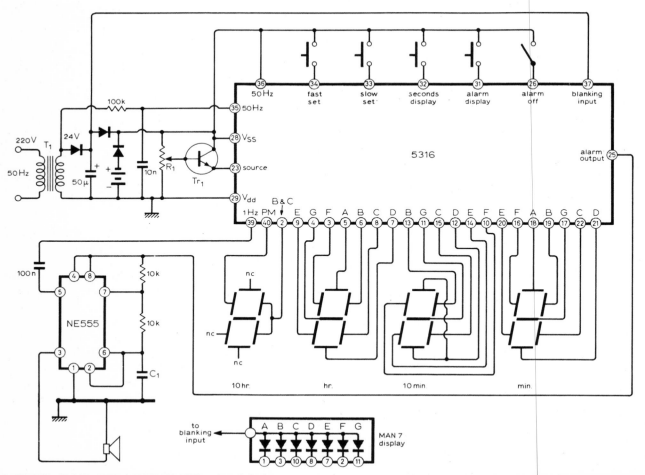

AC DIGITAL CLOCK WITH STANDBY BATTERY—Uses MM5316 alarm-clock IC, originally designed to drive LCD or fluorescent displays, but modified here for LED display. Diodes and batteries provide power if AC fails, with blanking of display to extend battery life. Accuracy is poor on batteries but batteries make resetting of time and alarm easier after AC interruption. Alarm uses 555 multivibrator to produce frequency-shift warble on output tone. Time is set by fast and slow buttons, and alarm is set with same buttons while depressing alarm-display button. Transistor type is not critical.—M. F. Smith, Digital Alarm Clock, *Wireless World,* Nov. 1976, p 62.

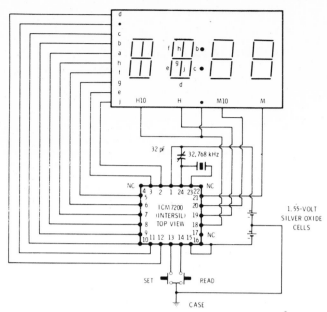

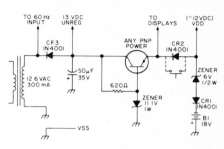

DIGITAL WRISTWATCH—Single Intersil ICM7200 IC drives multiplexed display giving choice of hours and minutes, seconds, and day/date. CMOS chip divides 32.768-kHz crystal output in long internal binary divider to produce basic 1-s clock rate. Further division gives other elements of display. Pressing read button once gives hours and minutes; pressing second time gives day and date; and pressing third time gives seconds.—D. Lancaster, "CMOS Cookbook," Howard W. Sams, Indianapolis, IN, 1977, p 377–378.

BATTERY BACKUP—During normal operation, all power for digital clock is provided by AC power supply. During power failure, clock continues operating from battery backup using two 9-V batteries in series. Battery drain is limited by diode CR2 that blocks power flow to displays. Optional switch may be installed across diode to short it for momentary viewing of display.—W. J. Prudhomme, CT7001 Clockbuster, *73 Magazine*, Dec. 1976, p 52–54 and 56–58.

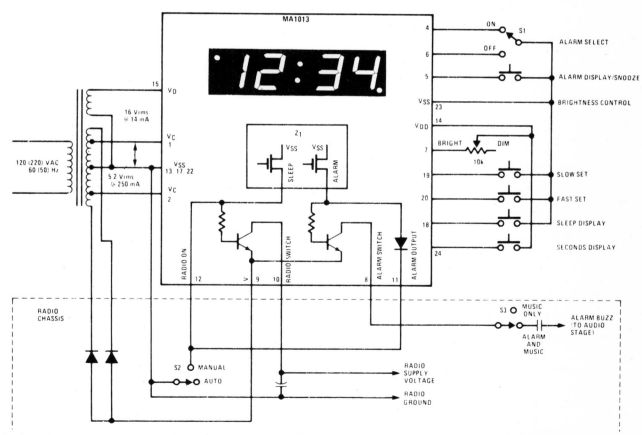

4-DIGIT 0.7-INCH LED DISPLAY—National MA1013 clock module contains MOS LSI clock IC, display, power supply, and associated discrete components on single printed-circuit board that is easily connected to radio. Operates from either 50-Hz or 60-Hz inputs, and gives either 12- or 24-h display format. Nonmultiplexed LED drive eliminates RF interference. Display is flashed at 1-Hz rate after power failure of any duration, to indicate need for resetting clock. Zero appearing in first digit is blanked. On 12-h version, dot in upper left corner is energized to indicate PM.—"MOS/LSI Databook," National Semiconductor, Santa Clara, CA, 1977, p 13-23–13-28.

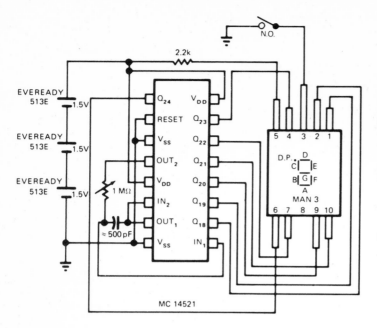

4-h DIGITAL WATCH—Single Motorola MC 14521 CMOS IC drives single-digit MAN 3 LED display in such a way that time range of 4 h is obtained with 1.875 min resolution. Can be built into old watch case at cost under $10 for parts. Oscillator frequency of 1.165 kHz can be tweaked to adjust clock, or crystal oscillator can be added for high accuracy. Analog/binary format of readout provides deciphering challenge to user, even though article gives diagram showing which segments of LED are lit for each time reading. Time intervals represented by each lit segment of display are: B = 2 h; C = 1 h; A = 30 min; F = 15 min; G = 7.5 min; E = 3.75 min; D = 1.875 min.—R. M. Steimle, Small CMOS Digital Watch Has Analog LED Output, *EDN Magazine,* Aug. 20, 1976, p 86.

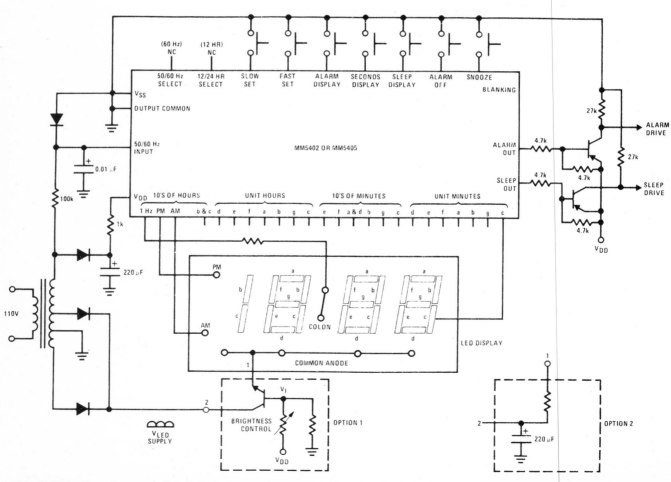

12-h ALARM—General-purpose digital clock with alarm uses National MM5402 or MM5405 MOS IC to drive 3½-digit LED display and provide drive for alarm. Brightness control is optional. Sleep output can be used to turn off radio after desired time interval of up to 59 min.—"MOS/LSI Databook," National Semiconductor, Santa Clara, CA, 1977, p 1-68–1-73.

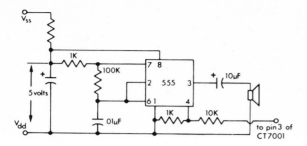

ALARM GENERATOR—Simple 555 timer generates alarm tone driving small loudspeaker, for use with Cal-Tex CT7001 and other similar digital clocks which do not have internal tone generator. Circuit requires +5 V, but supply can be higher value if suitable dropping resistor is used.—M. S. Robbins, "Electronic Clocks and Watches," Howard W. Sams, Indianapolis, IN, 1975, p 91.

0–9 s DIGITAL READOUT—Can be used for classroom demonstration of digital logic driving 7-segment LED or as attention-getting desk display. Time base Q1 feeds sequential timing pulses to 7490 decade counter. Pulses are counted in binary mode, and bit pattern corresponding to digits 0-9 is fed to 7447 binary-to-decimal decoder/driver connected to 7-segment readout. Calibrate with watch or with timing reference signals from WWV, adjusting R1 so display advances 1 digit per second.—F. M. Mims, "Electronic Circuitbook 5: LED Projects," Howard W. Sams, Indianapolis, IN, 1976, p 72–75.

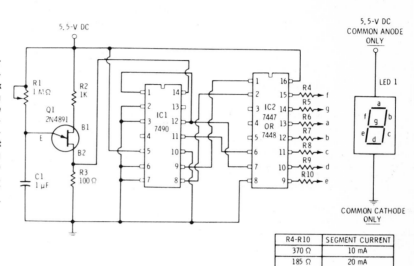

R4-R10	SEGMENT CURRENT
370 Ω	10 mA
185 Ω	20 mA

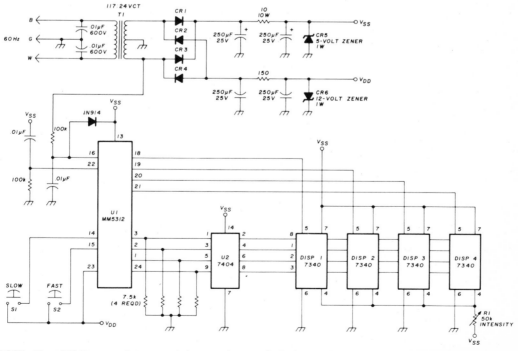

SIMPLE 24-h CLOCK—Use of 60-Hz power frequency as time base simplifies design while still giving long-term accuracy comparable to that of crystal time base. Four-digit display uses Hewlett-Packard 5082-7340 displays requiring only simple four-line BCD input. National MM5312N IC divides line frequency down to one pulse per minute and advances its internal storage register at same rate. Output of register is in binary form at pins 1, 2, 3, and 24, synchronized with digit-enable outputs at pins 18, 19, 20, and 21. Binary data is thus applied to all four displays in parallel, with enable lines controlling data feed. SN7404N inverter converts binary output data to TTL level required by displays. Power supply provides +5 V and −12 V for ICs and 60-Hz reference for clock check. CR5 is Radio Shack 276-561, CR6 is 276-563, and CR1-4 are 276-1146.—K. Powell, 24-Hour Clock with Digital Readout and Line-Frequency Time Base, *Ham Radio*, March 1977, p 44–48.

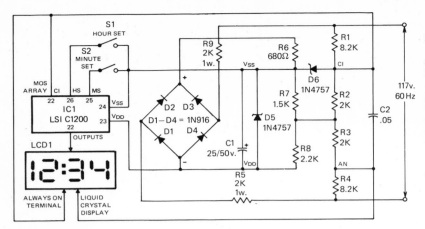

2-INCH LCD NUMERALS—Uses C1200 clock IC made by LSI Computer Systems, having time set, logic, division for seconds, minutes, and hours, 7-segment decoding, and display drivers and switches. Four-digit liquid crystal display panel (LCD) is MGC-50. S1 and S2 advance minutes or hours on display at 2-Hz rate for setting time. To use as elapsed-time indicator, close S1 and S2 simultaneously to generate reset pulse that sets timing change to zero. When both switches are released simultaneously, time count starts from zero.—R. F. Graf and G. J. Whalen, A Giant LCD Clock, *CQ*, Feb. 1978, p 18–23 and 76.

DIVIDE BY 5000 FOR CLOCK—Counter chain uses CD4017 that divides by integer from 2 to 10, selected by connecting appropriate output to reset. Extra gates recommended by RCA are not needed. Used in digital clock that changes automatically to battery operation when AC power fails. Clock operates on either 50 or 60 Hz.—S. I. Green, Digital Clock Keeps Counting Even When AC Power Fails, *EDN Magazine*, Dec. 20, 1974, p 49–51.

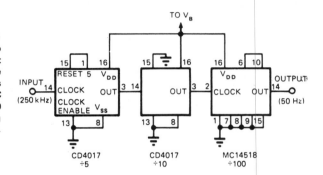

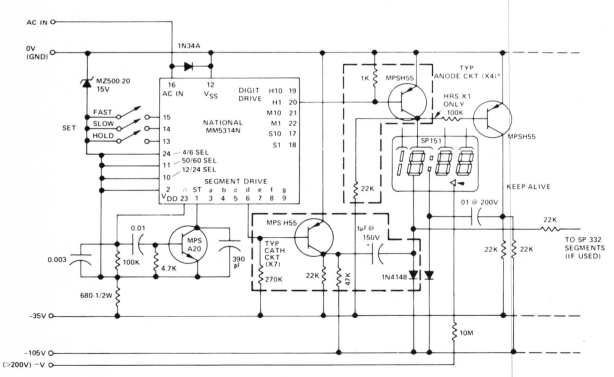

*(X4) FOR HRS, MINS, (X6) FOR HRS, MINS, SECS

4-DIGIT GAS DISPLAY—CMOS clock IC drives multidigit gas-discharge display. Simple circuit does not include alarm, flashing colon, and AM/PM features. Seven segment-driver circuits and four digit-driver circuits are required, although only one of each is shown. Additional drivers are needed if seconds display is desired. Required supply voltages can be obtained from transformer-type supply driving diode bridge; regulation is not needed.—M. S. Robbins, "Electronic Clocks and Watches," Howard W. Sams, Indianapolis, IN, 1975, p 68–71.

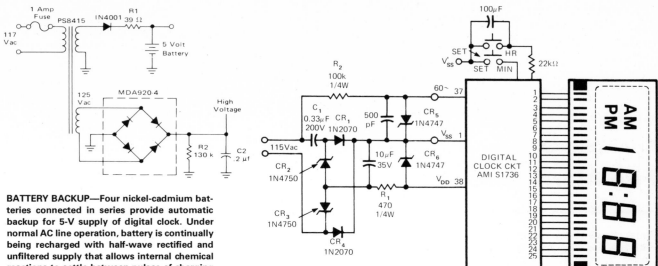

BATTERY BACKUP—Four nickel-cadmium batteries connected in series provide automatic backup for 5-V supply of digital clock. Under normal AC line operation, battery is continually being recharged with half-wave rectified and unfiltered supply that allows internal chemical reactions to settle between pulses of charging energy. R1 is chosen to make average charging current about 5% of battery rating.—D. Aldridge and A. Mouton, "Industrial Clock/Timer Featuring Back-Up Power Supply Operation," Motorola, Phoenix, AZ, 1974, AN-718A, p 7.

12- OR 24-h CLOCK—Single American Microsystems AMI S1736 clock chip drives liquid-crystal readout to give either 12-h display with AM/PM indicator or 24-h digital display by changing only three connections.—LSI in Consumer Applications, Round 2: Clocks on a Chip, *EDN Magazine,* May 5, 1973, p 22–23.

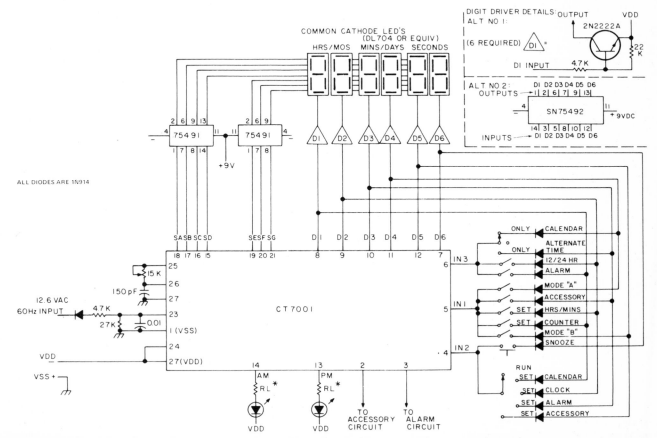

CALENDAR CLOCK—Uses Fairchild FCM7001 IC equivalent of Cal-Tex CT7001 clock chip (which is no longer available) to drive six 7-segment LEDs that can be switched to show 12- or 24-h time and 28/30/31 calendar, along with alarm features. Article gives construction details. Each SN75491 driver chip has pins 3, 5, 10, and 12 connected to pin 11 through 150-ohm resistor. RL is typically 2.7K, chosen to limit LED current to less than 5 mA.—W. J. Prudhomme, CT7001 Clockbuster, *73 Magazine,* Dec. 1976, p 52–54 and 56–58.

CHAPTER 5
Fire Alarm Circuits

Sensors used may respond to gas, ionization, flame, or smoke associated with fire, for triggering circuits driving variety of alarm devices.

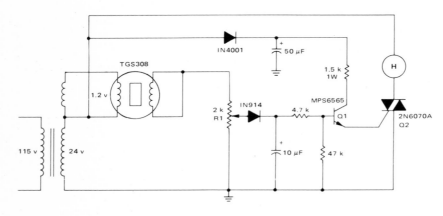

TRIAC GAS/SMOKE DETECTOR—Conductivity of Taguchi TGS308 gas sensor increases in presence of combustible gases, increasing load voltage across R1 from normal 3 VRMS to as much as 20 V. Rise in voltage trips comparator to turn on transistor Q1 that supplies trigger current to 2N6070A sensitive-gate triac. Resulting full-wave drive of Delta 16003168 24-VAC horn gives sound output of 90 dB at 10 feet. Horn stops automatically when gas clears sensor.—A. Pshaenich, "Solid State Gas/Smoke Detector Systems," Motorola, Phoenix, AZ, 1975, AN-735, p 4.

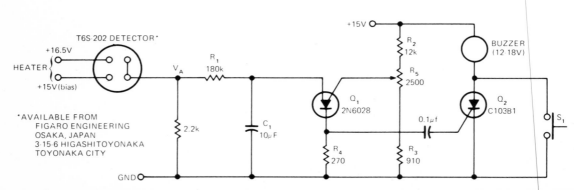

GAS/SMOKE SENSOR—Sensor is based on selective absorption of hydrocarbons by N-type metal-oxide surface. Heater in sensor burns off hydrocarbons when gas or smoke disappears, to make sensor reusable. Requires initial warm-up time of about 15 min in hydrocarbon-free environment. When gas or smoke is present, V_A quickly rises and triggers programmable UJT Q_1. Resulting voltage pulse across R_4 triggers Q_2 and thereby energizes buzzer. S_1 is reset switch. R_1 and C_1 give time delay that prevents triggering by small transients such as smoke from cigarette. R_5 adjusts alarm threshold. Use regulated supply.—S. J. Bepko, Gas/Smoke Detector Is Sensitive and Inexpensive, EDN Magazine, Sept. 20, 1973, p 83 and 85.

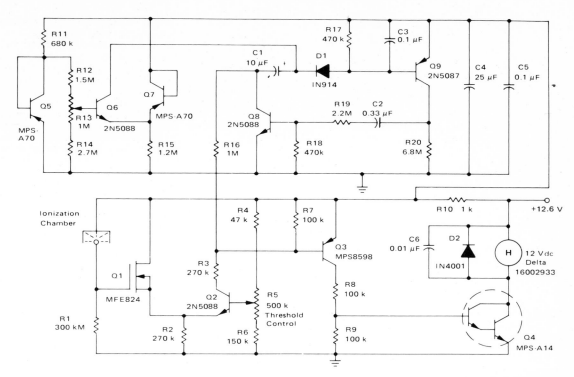

IONIZATION ALARM USING TRANSISTORS— Use of continuous smoke alarm signal rather than beeping horn simplifies transistor circuits needed to trigger fire alarm and low-battery alarm. When high impedance of ionization chamber is lowered by smoke or gas, amplifier Q1-Q2-Q3 supplies 100-μA base current to Darlington Q4 for powering horn continuously as long as smoke content exceeds that set by threshold control R5. Low-battery circuit is tripped at voltage range between 9.8 and 11.2 V, as determined by R13, to energize MVBR Q8-Q9 for driving horn 0.7 s, with 50-s OFF intervals. Battery is chosen to last at least 1 year while furnishing standby current of about 70 μA.—A. Pshaenich, "Solid State Gas/Smoke Detector Systems," Motorola, Phoenix, AZ, 1975, AN-735, p 8.

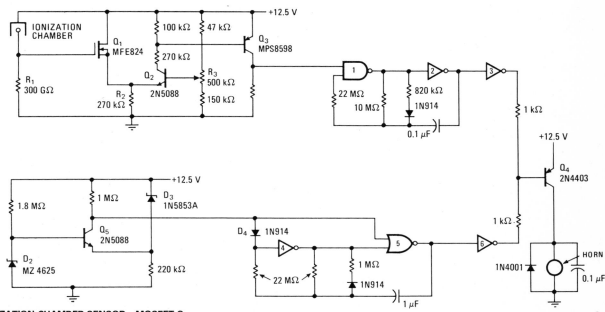

IONIZATION-CHAMBER SENSOR—MOSFET Q1 with high input impedance monitors voltage level at divider formed by R1 and ionization chamber, with output of Q1 going to Q2 which forms other half of differential amplifier. With smoke level of 2% or higher, Q3 is turned on and applies logic 1 to one input of NAND gate 1 in asymmetrical astable MVBR. Capacitor in MVBR charges quickly and discharges slowly, making alarm horn sound during discharge via inverter 3 and driver transistor Q4. Comparator circuit Q5 drives second MVBR to energize horn through inverter 6 and same driver Q4 when battery is low, but with distinctive 1-s toot every 23 s to conserve energy remaining in battery and differentiate from fire warning.—A. Pshaenich and R. Janikowski, Gas and Smoke Detector Uses Low-Leakage MOS Transistor, *Electronics,* Nov. 28, 1974, p 124–125.

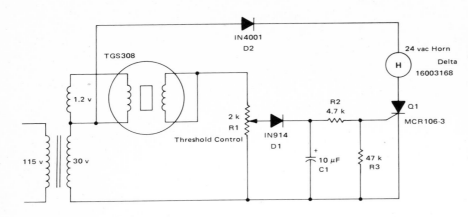

SCR GAS/SMOKE DETECTOR—Simple circuit uses Taguchi TGS308 gas sensor with SCR Q1 for half-wave control of 24-VAC alarm horn. Sensor is based on adsorptive and desorptive reaction of gases on tin oxide semiconductor surface encased in noble-metal heater that serves also as electrode. Combustible gases increase conductivity of sensor, thereby increasing load voltage enough to trip comparator and initiate alarm. Output voltage across R1 is normally about 3 VRMS. With gas or smoke, voltage can rise to 20 V. When gas or smoke has cleared sensor, SCR turns off at first zero crossing. Drawbacks are absence of time delay for preventing false alarm when power is turned on and reduced sound level of horn with half-wave operation.—A. Pshaenich, "Solid State Gas/ Smoke Detector Systems," Motorola, Phoenix, AZ, 1975, AN-735, p 3.

FLAME DETECTOR DRIVES TTL LOAD—Sensor is silicon Darlington phototransistor Q_1 having peak response near infrared bands. Filter is required to reduce interference from visible light sources. Circuit is sensitive enough to pick up hydrogen flames that emit no visible light. Article describes operation of circuit and gives design equations. Output can go directly to input port of microprocessor.—A. Ames, This Flame Detector Interfaces Directly to a µP, *EDN Magazine,* Oct. 20, 1976, p 122 and 124.

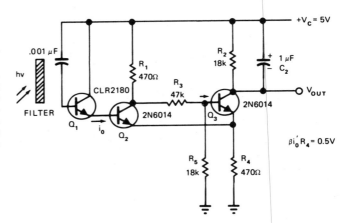

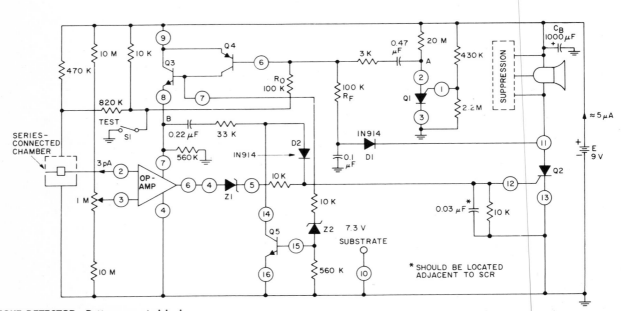

SMOKE DETECTOR—Battery-operated ionization-type smoke detector uses RCA CA3130 opamp as interface for ionization chamber that provides picoampere currents. With opamp in pulsed mode (on for 20 ms of 20-s period), IC draws only 0.6 µA average instead of 600 µA. Other active components and zener, all on RCA CA3097 array, provide low-battery monitor and horn-driver functions. When chamber detects smoke, combination of R_F and D1 provides sufficient base current to keep Q3 and Q4 on. Opamp is then powered continuously, and steering diode Z1 supplies continuous current to gate of Q2 for energizing horn. Battery drain is only 5 mA in monitoring mode.—G. J. Granieri, Bipolar-MOS and Bipolar IC's Building Blocks for Smoke-Detector Circuits, *IEEE Transactions on Consumer Electronics,* Nov. 1977, p 522–527.

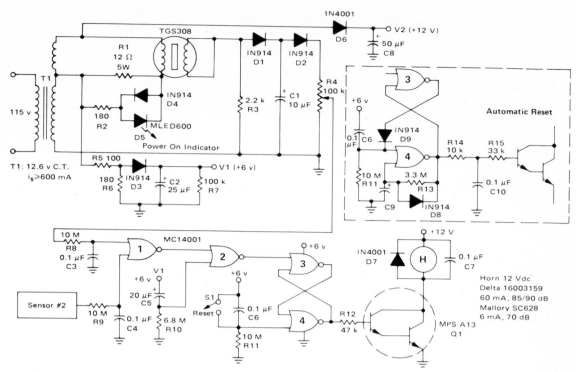

GAS/SMOKE DETECTOR WITH LATCH—CMOS latching logic provides 2-min time delay to prevent false alarm when power is first applied to fire alarm using Taguchi TGS308 gas sensor whose conductivity increases in presence of combustible gases. Normal voltage of 3 VRMS across R4 increases to about 20 V in presence of fire. Half of 12.6-V center-tapped transformer secondary is used for 6-V supply and full 12.6 V for DC horn supply. Latch is reset manually with S1 to turn off alarm after gas level drops. Optional circuit shown can be used for automatic reset.—A. Pshaenich, "Solid State Gas/Smoke Detector Systems," Motorola, Phoenix, AZ, 1975, AN-735, p 5.

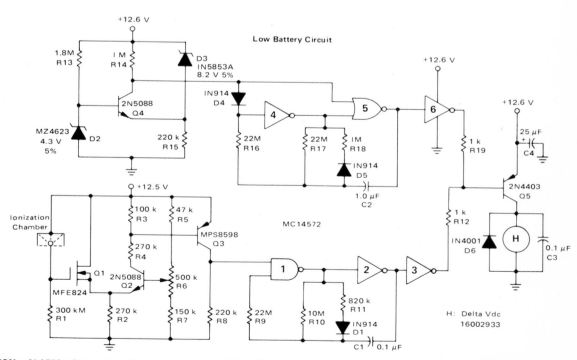

IONIZATION ALARM—Gates in Motorola MC14572 CMOS IC form two alarm oscillators, one energized in presence of smoke at ionization chamber and other for low battery. Standby currents of circuits are low enough to give at least 1 year of operation from 750-mAh battery. R6 is adjusted to give desired smoke detection sensitivity. Gates 1 and 2 form MVBR that drives horn at astable rate of 2.5 s on and 0.2 s off in presence of smoke. When battery is low, comparator Q4-D2-D3 trips (about 10.5 V) and energizes inverter 4 of low-battery astable MVBR. DC horn is then powered at astable rate of about 1 s every 23 s to give early warning of need to change battery.—A. Pshaenich, "Solid State Gas/Smoke Detector Systems," Motorola, Phoenix, AZ, 1975, AN-735, p 7.

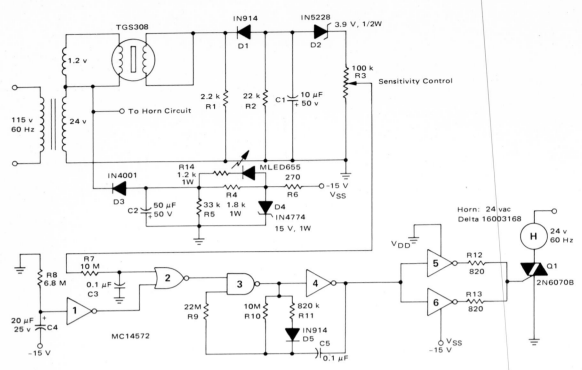

GAS/SMOKE DETECTOR WITH BEEPING HORN—Taguchi TGS308 gas sensor increases voltage across R3 when sensor conductivity is increased by combustible gases. After time delay provided to prevent power turn-on false alarms, CMOS astable MVBR using gates 3 and 4 is energized to fire triac and drive AC horn to give distinctive repetitive sound lasting about 2.5 s, with 0.2-s intervals between beeps. Triac gate drivers operate from −15 V supply derived from 24-V winding of power transformer.—A. Pshaenich, "Solid State Gas/Smoke Detector Systems," Motorola, Phoenix, AZ, 1975, AN-735, p 6.

CHAPTER **6**
Flasher Circuits

Provide fixed or variable flash rates for LEDs, incandescent lamps, or fluorescent lamps used as indicators, alarms, warnings, and for such special effects as Christmas-light shimmer. See also Game and Lamp Control chapters.

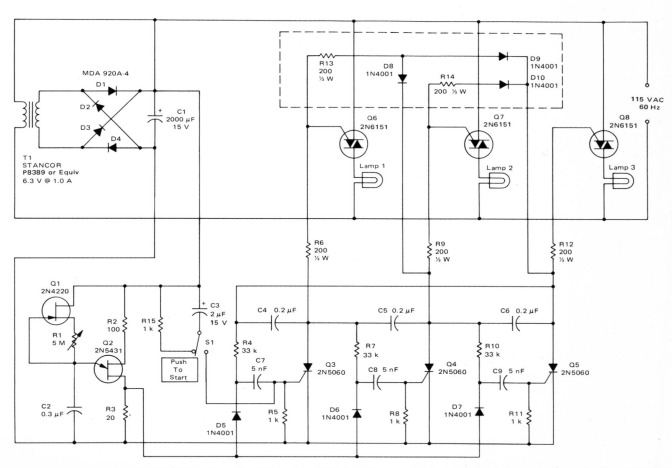

SEQUENTIAL AC FLASHER—Uses simple ring counter in which triac gates form part of counter load. Incandescent lamps come on in sequence, with only one lamp normally on at a time. Pulse rate for switching lamp can be adjusted from about 1 every 0.1 s to 1 every 8 s. Circuit enclosed in dashed rectangle can be added to keep previous lamps on when next lamp is turned on. Only three stages are shown, but any number of additional stages can be added.—"Circuit Applications for the Triac," Motorola, Phoenix, AZ, 1971, AN-466, p 11.

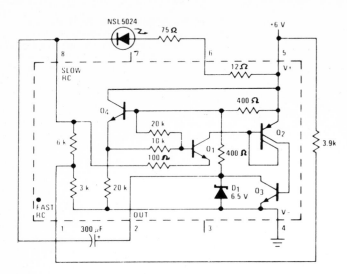

VARIABLE FLASHER FOR LED—Terminal connections of National LM3909 flasher IC give choice of three different flash rates for LED used as indicator in battery portable equipment. External resistors provide additional adjustments of flash rate. Appropriate connections to pins 1 and 8 make flash-controlling internal resistance 3K, 6K, or 9K. Flasher operates at any supply voltage above 2 V, with low duty cycle to give long battery life.—P. Lefferts, Power-Miser Flasher IC Has Many Novel Applications, *EDN Magazine*, March 20, 1976, p 59–66.

SHIMMER FOR CHRISTMAS LIGHTS—Circuit uses half of AC cycle to power lights conventionally. On other half-cycle, C charges and builds up voltage on gate of SCR. When firing point is reached, SCR conducts and allows remainder of this half-cycle to pass through light string. Result is flash that gives shimmer or strobe effect. C is 100-μF 50-V electrolytic, R1 is 2.7K, R2 is 22K, R3 is 3.3K, R4 is 100K pot, and R5 is 1K. Diodes are Motorola HEP R0053. SCR is GE C106B1 or Motorola HEP R1221 mounted on heatsink.—R. F. Graf and G. J. Whalen, Add Shimmer to Your Christmas Lights, *Popular Science*, Dec. 1973, p 124.

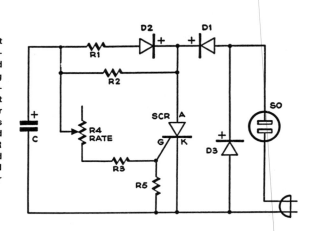

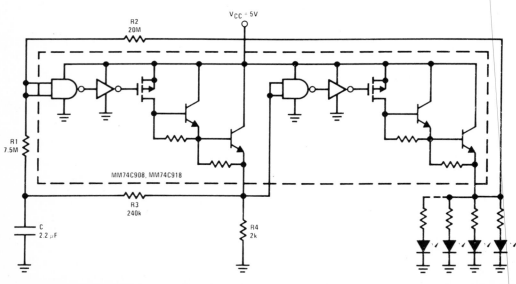

DRIVING LED ARRAY—National MM74C908/ MM74C918 dual CMOS driver has sections connected as Schmitt-trigger oscillator, with R1 and R2 used to generate hysteresis. R3 and C are inverting feedback timing elements, and R4 is pulldown load for first driver. Output current drive capability is greater than 250 mA, making circuit suitable for driving array of LEDs or lamps.—"CMOS Databook," National Semiconductor, Santa Clara, CA, 1977, p 5-38–5-49.

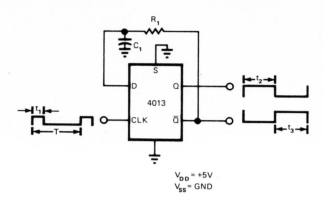

V_{DD} = +5V
V_{SS} = GND

1-Hz LAMP BLINKER—Single CMOS flip-flop generates approximately constant low-frequency signal from variable high-frequency signal. RC network in feedback loop determines output frequency, which is independent of rate at which flip-flop is clocked if output frequency is lower than clock frequency. If clock frequency is lower, output transitions occur at half of clock frequency. Provides two outputs, approximately equal in duty cycle but opposite in phase. Circuit was developed to blink lamp at 1 Hz to indicate presence of active digital signal having variable duty cycle in range of 100 to 3000 Hz.—V. L. Schuck, Generate a Constant Frequency Cheaply, *EDN Magazine,* Aug. 20, 1975, p 80 and 82.

3-V STROBE—Flash rate of 1767 lamp can be adjusted from no flashes to continuously on, in circuit using National LM3909 flasher IC with external NPN power transistor rated at 1 A or higher. Can be used as variable-rate warning light, for advertising, or for special effects. With lamp in large reflector in dark room, flashes several times per second are almost fast enough to stop motion of dancer.—P. Lefferts, Power-Miser Flasher IC Has Many Novel Applications, *EDN Magazine,* March 20, 1976, p 59–66.

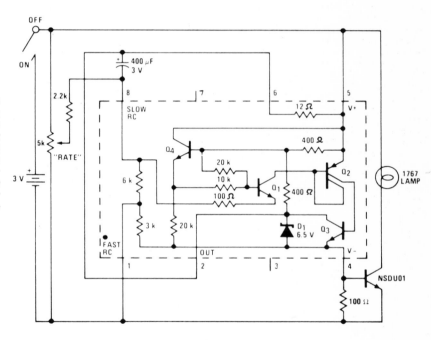

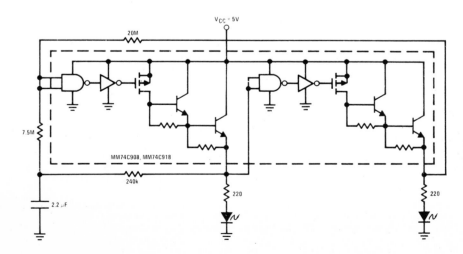

OUT-OF-PHASE DOUBLE FLASHER—Sections of National MM74C908/MM74C918 dual CMOS driver are connected as Schmitt-trigger oscillator, with LEDs at output of each section so LEDs will flash 180° out of phase. High output current capability makes circuit suitable for driving two LED arrays.—"CMOS Databook," National Semiconductor, Santa Clara, CA, 1977, p 5-38–5-49.

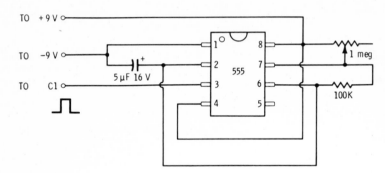

CLOCK DRIVE FOR FLIP-FLOP FLASHER—555 timer connected as astable MVBR generates series of timing pulses at rate determined by value of capacitor and setting of 1-megohm pot. Provides automatic string of input pulses for driving flip-flop of dual flasher. Pulse output goes to input capacitor C1 of flip-flop.—F. M. Mims, "Integrated Circuit Projects, Vol. 5," Radio Shack, Fort Worth, TX, 1977, 2nd Ed., p 30–37.

SEQUENTIAL SWITCHING OF LOADS—Ring counter using four-layer diodes D_N provides sequential switching of loads under control of input pulse-train signal. Indicator lamps are shown, but any load from 15 to 200 mA can be switched. After power is applied, reset switch must be pressed to establish current through L. When switch is released, this current flows through C_2 and breaks down D_2, allowing current to flow through first lamp I_1. Input pulse to transistor Q (normally held off by current through R_1) turns Q off and removes power from diode circuits, thus turning I_1 and D_2 off. At end of input pulse, Q comes on and restores power to diode circuits, but all loads will be turned off. Voltage on C_3 now adds to 6 V normally across D_4, making D_4 break down and turn on I_2. Next input pulse will break down D_6 in same manner. Output signals may be picked up as negative pulses at A or B or by current-sensing at C if required for controlling larger loads.—J. Bliss and D. Zinder, "4-Layer and Current-Limiter Diodes Reduce Circuit Cost and Complexity," Motorola, Phoenix, AZ, 1974, AN-221, p 5.

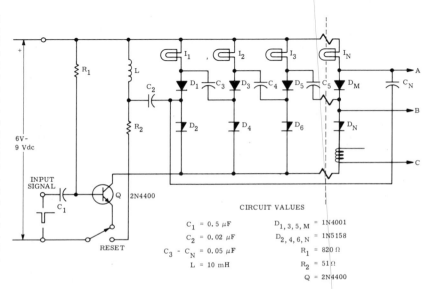

CIRCUIT VALUES

$C_1 = 0.5\ \mu F$ $D_{1,3,5,M} = 1N4001$

$C_2 = 0.02\ \mu F$ $D_{2,4,6,N} = 1N5158$

$C_3 - C_N = 0.05\ \mu F$ $R_1 = 820\ \Omega$

$L = 10\ mH$ $R_2 = 51\ \Omega$

$Q = 2N4400$

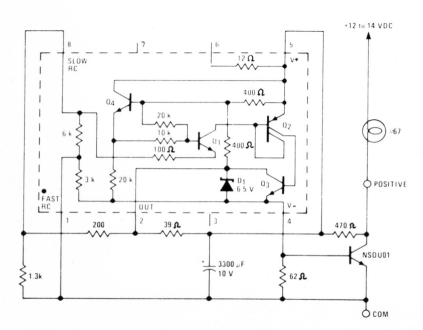

1-Hz AUTO FLASHER—Lamp drawing nominal 600 mA is flashed at 1 Hz by National LM3909 flasher IC operating from 12-V automotive battery. Use of 3300-μF capacitor makes flasher IC immune to supply spikes and provides means of limiting IC supply voltage to about 7 V.—P. Lefferts, Power-Miser Flasher IC Has Many Novel Applications, *EDN Magazine,* March 20, 1976, p 59–66.

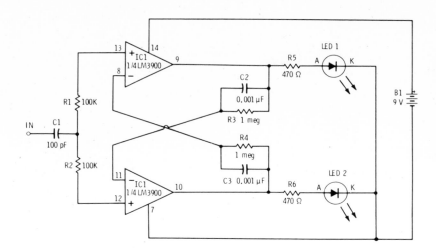

DEMONSTRATION FLIP-FLOP—Two sections of LM3900 quad opamp form bistable MVBR for flip-flop having two stable states. When input is grounded momentarily, output of one of opamps swings completely on and turns other opamp off. LED indicates which opamp is on at any particular time. Next grounding of input reverses conditions. Ideal for classroom demonstrations.—F. M. Mims, "Integrated Circuit Projects, Vol. 5," Radio Shack, Fort Worth, TX, 1977, 2nd Ed., p 30–37.

SCR FLASHES LED—UJT oscillator Q1 provides timing pulses for triggering SCR driving red Radio Shack 276-041 LED. Circuit draws only 2 mA from 9-V battery when producing 12 flashes per second. SCR is 6-A 50-V 276-1089.—F. M. Mims, "Semiconductor Projects, Vol. 2," Radio Shack, Fort Worth, TX, 1976, p 78–84.

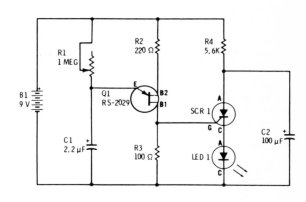

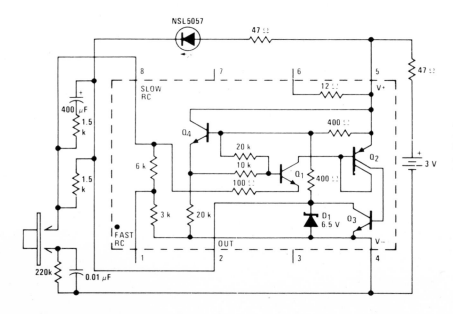

SINGLE-FLASH LED—Mono MVBR connection of National LM3909 IC produces 0.5-s flash with LED each time pushbutton makes momentary contact.—"Linear Applications, Vol. 2," National Semiconductor, Santa Clara, CA, 1976, AN-154, p 9.

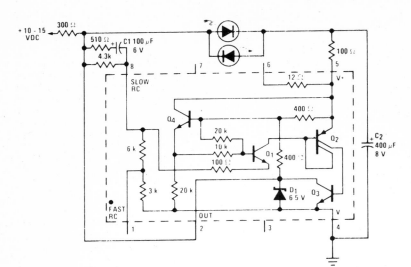

ALTERNATING RED/GREEN—National LM3909 IC is connected as relaxation oscillator for flashing red and green LEDs alternately. With 12-VDC supply, repetition rate is about 2.5 Hz. Green LED should have its anode or positive lead toward pin 5 as shown for lower LED, where shorter but higher-voltage pulse is available. LED types are not critical.—"Linear Applications, Vol. 2," National Semiconductor, Santa Clara, CA, 1976, AN-154, p 3.

ALARM-DRIVEN FLASHER—Simple two-transistor flasher circuit for annunciator system is activated by alarm. Operator acknowledges alarm condition by depressing S_A, which changes lamp from flashing to steady ON condition. 6-V incandescent lamp draws about 0.3 A through Q_2, but 1K load resistor for Q_1 limits current of this transistor to about 6 mA so smaller transistor can be used.—T. Stehney, Flasher Design Cuts Extra Components, *EDN Magazine*, Sept. 20, 1978, p 144.

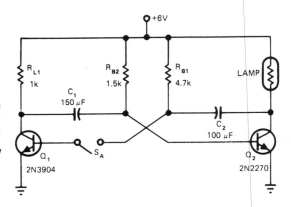

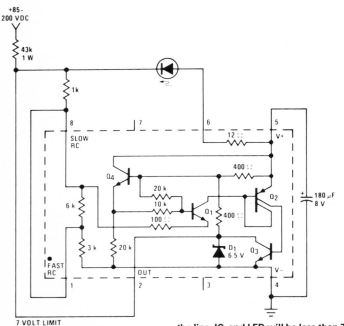

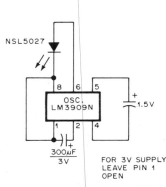

FLASHING LED IS REMOTE MONITOR—Circuit uses National LM3909 flasher IC to drive LED for monitoring remotely located high-voltage power supply. When 43K dropping resistor is located at power supply, all other voltages on the line, IC, and LED will be less than 7 V above ground, for safe remote monitoring. Use any LED drawing less than 150 mA.—P. Lefferts, Power-Miser Flasher IC Has Many Novel Applications, *EDN Magazine*, March 20, 1976, p 59–66.

1.5-V OR 3-V INDICATOR—Digi-Key LM3909N flasher/oscillator drives LED serving as ON/OFF indicator for battery-operated devices. At flash rate of 2 Hz, battery life almost equals shelf life.—C. Shaw, ON-OFF Indicator for Battery Device, *QST*, March 1978, p 41–42.

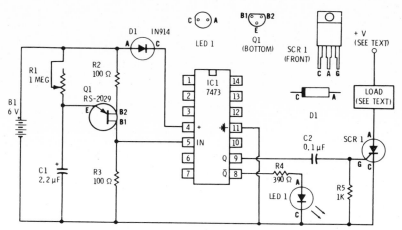

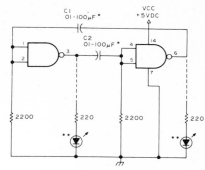

FLIP-FLOP DRIVES SCR—UJT relaxation oscillator Q1 serves as clock for driving section of 7473 dual flip-flop. One output of flip-flop flashes Radio Shack 276-041 red LED to indicate operating status. Other output alternately triggers SCR which can be 6-A 50-V 276-1089, for flashing lamp load. Load and SCR supply voltage depend on application but must be within SCR rating.—F. M. Mims, "Semiconductor Projects, Vol. 2," Radio Shack, Fort Worth, TX, 1976, p 62–70.

LED BLINKER—Two sections of SN7400 quad gate form MVBR operating at low enough frequency so LED status indicators come on and off slowly for visual observation of MVBR. LEDs are optional and do not affect operation of MVBR. Capacitors must be same value. Ideal for student demonstration in classroom or as Science Fair exhibit.—A. MacLean, How Do You Use ICs?, *73 Magazine*, Dec. 1977, p 56–59.

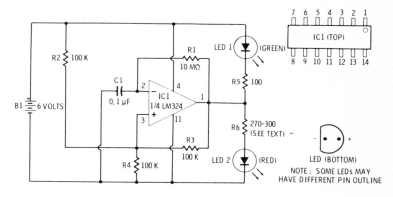

RED/GREEN LED FLASHER—One section of LM324 quad opamp is connected as square-wave generator giving about 1 flash per second for each LED. Series resistors for LEDs have different values because they have different forward voltage requirements. If LED 2 glows between flashes, increase value of R6 slightly. Too large a value for R6 reduces flash brilliance of LED 2. Supply can be 5 or 6 V.—F. M. Mims, "Semiconductor Projects, Vol. 1," Radio Shack, Fort Worth, TX, 1975, p 69–74.

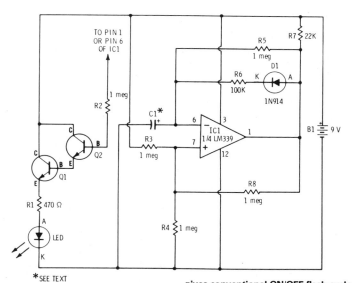

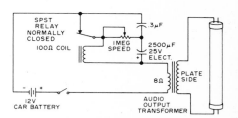

COMPARATOR LED FLASHER—One section of LM339 quad comparator drives two RS2016 NPN transistors having LED load, to give simple flasher for classroom demonstrations. Circuit can be duplicated with other three sections to give four flashers. Connecting R2 to pin 1 of IC gives conventional ON/OFF flash cycle in which LED turns on and off rapidly. Connecting R2 to pin 6 makes LED turn on rapidly and turn off very slowly. C1 controls flash interval; typical value is 0.01 μF.—F. M. Mims, "Integrated Circuit Projects, Vol. 5," Radio Shack, Fort Worth, TX, 1977, 2nd Ed., p 45–51.

12-V FLUORESCENT—Relay acts as mechanical DC/AC converter operating off 12-V car battery. Each time relay opens, inductive kick in relay coil is stepped up by output transformer to high enough voltage for ionizing 24-inch fluorescent tube, giving flash that can serve as emergency flasher when car breaks down.—Circuits, *73 Magazine*, June 1975, p 175.

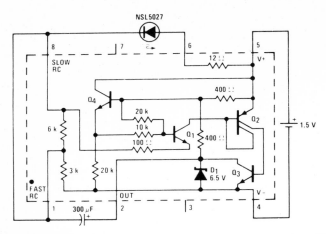

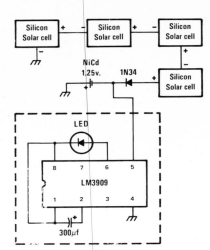

1.5-V LED FLASHER—National LM3909 IC operating from 1.5-V battery drives NSL5027 LED in such a way that current is drawn by LED only about 1% of time. External 300-μF capacitor sets flash rate at about 1 Hz.—"Linear Applications, Vol. 2," National Semiconductor, Santa Clara, CA, 1976, AN-154, p 2.

LED FLASHER—Requires only LM3909 IC and external capacitor operating from 1.25-V nicad or other penlight cell. Circuit can be duplicated for as many additional flashing LEDs as are desired for display. Optional charging circuit uses silicon solar cells and diode for daytime charging of battery automatically.—J. A. Sandler, 11 Projects under $11, *Modern Electronics*, June 1978, p 54–58.

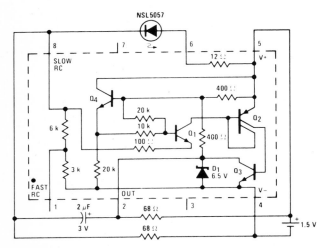

2-kHz FLASHER FOR LED—Single 1.5-V cell provides power for National LM3909 flasher IC that operates at high enough frequency to appear on continuously, for use as indicator in battery portable equipment. Duty cycle and frequency of current pulses to LED are increased by changing external resistors until average energy reaching LED provides sufficient light for application. At 2 kHz, no flicker is noticeable.—P. Lefferts, Power-Miser Flasher IC Has Many Novel Applications, *EDN Magazine*, March 20, 1976, p 59–66.

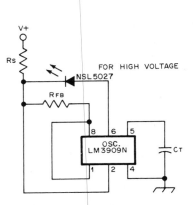

6-V OR 15-V INDICATOR—Uses Digi-Key LM3909N flasher/oscillator to drive LED at 2 Hz as ON/OFF indicator for battery-operated devices. For 6-V battery, C_T is 400 μF, R_S is 1000 ohms, and R_{FB} is 1500 ohms. For 15 V, corresponding values are 180, 3900, and 1000. Battery life is essentially same as shelf life.—C. Shaw, ON-OFF Indicator for Battery Device, *QST*, March 1978, p 41–42.

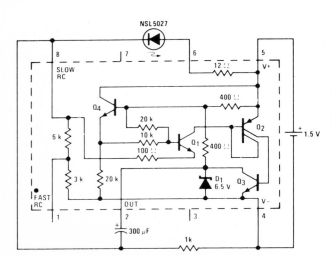

FAST 1.5-V BLINKER—Addition of 1K resistor between pins 4 and 8 of National LM3909 IC increases flash rate to about 3 times that obtainable when 300 μF is connected between pins 1 and 2. Modification of external connections gives choice of 3K, 6K, or 9K for internal RC resistors.—"Linear Applications, Vol. 2," National Semiconductor, Santa Clara, CA, 1976, AN-154, p 2.

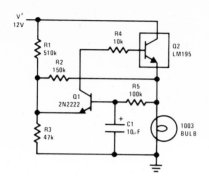

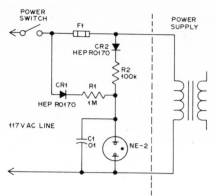

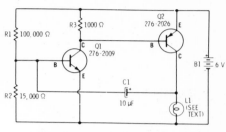

1-A LAMP FLASHER—National LM195 power transistor is turned on and off once per second for flashing 12-V lamp. Current limiting in LM195 prevents high peak currents during turn-on even though cold lamp can draw 8 times normal operating current. Current-limiting feature prolongs lamp life in flashing applications.—R. Dobkin, "Fast IC Power Transistor with Thermal Protection," National Semiconductor, Santa Clara, CA, 1974, AN-110, p 5.

BLOWN-FUSE BLINKER—Neon lamp NE-2 glows steadily when fuse is good and flashes when fuse opens. Flash rate, determined by R1 and C1, is about 10 flashes per second for values shown.—T. Lincoln, A "Smart" Blown-Fuse Indicator, *QST*, March 1977, p 48.

AUTO-BREAKDOWN FLASHER—Two-transistor amplifer with regenerative feedback sends 60-ms pulses of currents up to several amperes through low-voltage lamp to give high-brilliance flashes without destroying lamp. L1 can be PR-2 lamp (Radio Shack 272-1120).—F. M. Mims, "Transistor Projects, Vol. 1," Radio Shack, Fort Worth, TX, 1977, 2nd Ed., p 27–32.

CHAPTER 7
Game Circuits

Included are chip connections, VHF modulators, score generators, and sound effects for variety of TV games, along with electronic dice, roulette wheel, coin tosser, robot toy, model railroad switch, six-note chimes, and attention-getting LED displays.

RIFLE—Developed for use with General Instruments AY-3-8500-1 TV game chip to simulate target practice with rifle. Player aims at bright target spot moving randomly across TV screen. If gun is on target when trigger is pulled, phototransistor in barrel picks up light from target and generates pulse for producing sound effect of hit and incrementing player's score. PT-1 can be TIL64 or equivalent phototransistor. 4098 is dual mono, and 4011 is quad two-input NAND gate. Pulse outputs go to pins of game chip. Article gives all circuits but covers construction only in general terms.—S. Ciarcia, Hey, Look What My Daddy Built!, *73 Magazine*, Oct. 1976, p 104–108.

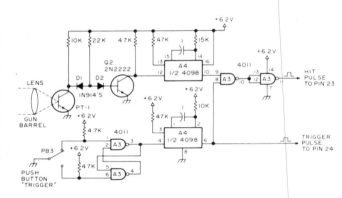

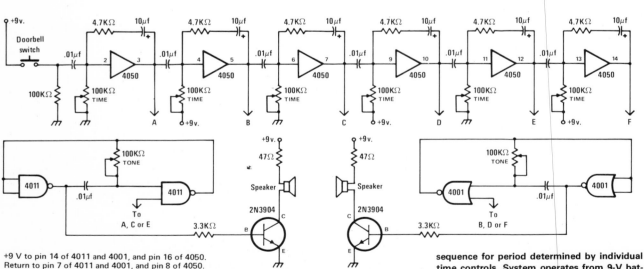

+9 V to pin 14 of 4011 and 4001, and pin 16 of 4050. Return to pin 7 of 4011 and 4001, and pin 8 of 4050.

SIX-TONE CHIME—Separate AF oscillators, gated on by six-stage time-delay circuit, generate six different chime tones. Loudspeakers can be mounted so each tone comes from different location in house. When doorbell button is pushed, each tone generator is turned on in sequence for period determined by individual time controls. System operates from 9-V battery, with CMOS logic drawing very little standby current.—J. Sandler, 9 Projects under $9, *Modern Electronics*, Sept. 1978, p 35–39.

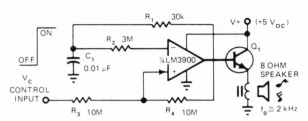

DIGITAL NOISEMAKER—Simple sound-effect generator for video games, electronic cash registers, and electronic toys uses one-fourth of LM3900 quad opamp chip as 2-kHz signal generator that can be turned on or off by input control voltage. Suitable for applications that do not require pure sine wave. Output transistor Q_1, needed with low-impedance voice coil, is not critical as to type. For smaller acoustic output, Q_1 can be replaced by 100-ohm resistor if 100-ohm voice coil is used, to avoid overloading IC.—T. Frederiksen, Build a Transformerless Tone Annunciator, *EDN Magazine*, April 5, 1977, p 141–142.

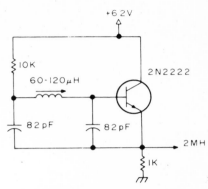

2-MHz MASTER CLOCK—Developed for use with General Instruments AY-3-8500-1 TV game chip, which contains dividers that deliver required 60-Hz vertical and 15.75-kHz horizontal sync signals for video signal going to TV set. Coil is Miller 9055 miniature slug-tuned. Article gives other circuits for game.—S. Ciarcia, Hey, Look What My Daddy Built!, *73 Magazine*, Oct. 1976, p 104–108.

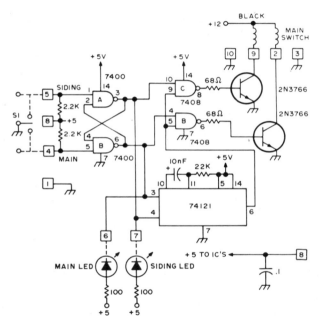

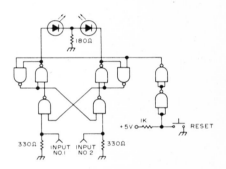

WHO'S FIRST?—One of LEDs comes on to indicate which of two people pushes button first after event such as stopping of music. Circuit requires two 7400 quad gates.—Circuits, *73 Magazine*, Nov. 1974, p 142.

1, 3, 10	GROUND	
2	TO SWITCH COIL—MAIN	SEPARATE + 12V SUPPLY 1
9	TO SWITCH COIL—SIDING	AMP, (UNREGULATED) USED FOR SWITCHES
4	CONTROL—SHORT TO GROUND TO THROW SWITCH TO MAIN LINE	
5	CONTROL—SHORT TO GROUND TO THROW SWITCH TO SIDING	
6	LED TO +5 TO INDICATE SWITCH IN MAIN (THIS POINT LOW)	
7	LED TO +5 TO INDICATE SWITCH IN SIDING	
8	+5 VOLTS IN FOR ICs	

POINTS 4 AND 5 CAN BE PARALLEL TO MANUAL MOMENTARY SWITCHES AND LOGIC SWITCHES—ANY PULSE (LOW) WILL WORK, HOLDING POWER ON ABOUT ½ SECOND, 74121 WITH RESISTOR AND CAP CONTROL TIME.

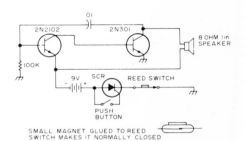

HOWLING BOX—Tone oscillator driving loudspeaker is sealed into wood or plastic box, with reed switch mounted on one face of box and pushbutton of other switch projecting out through hole in box. Place "DO NOT TOUCH" label on button. When button is pushed despite warning, SCR latches and applies power to AF oscillator. Only way to turn off howling is to hold large permanent magnet against location of reed switch, to oppose field of magnet glued on switch and make reed contacts open. If mercury switch is used in box in place of pushbutton, alarm goes off when box is picked up.—P. Walton, Now What Have I Done?, *73 Magazine*, May 1975, p 81.

MODEL RAILROAD SWITCHING—Control circuit is used to drive solenoid-operated track switches of typical HO train layout. Input can be pair of complementary TTL signals from 8008 or other computer or can be from manual switch S1. 74121 mono MVBR controls time that switch is energized in given direction. Output transistors are rated at 20 W, enough for driving solenoids taking 1 A at 12 V. Use protective diodes across coils of solenoids.—H. De Monstoy, Model Railroad Switch Control Circuit, *BYTE*, Oct. 1975, p 87.

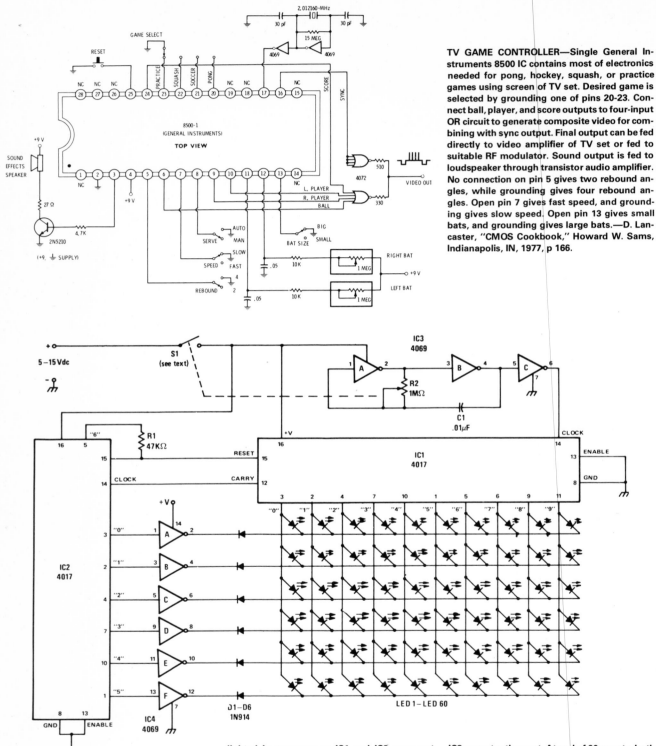

TV GAME CONTROLLER—Single General Instruments 8500 IC contains most of electronics needed for pong, hockey, squash, or practice games using screen of TV set. Desired game is selected by grounding one of pins 20-23. Connect ball, player, and score outputs to four-input OR circuit to generate composite video for combining with sync output. Final output can be fed directly to video amplifier of TV set or fed to suitable RF modulator. Sound output is fed to loudspeaker through transistor audio amplifier. No connection on pin 5 gives two rebound angles, while grounding gives four rebound angles. Open pin 7 gives fast speed, and grounding gives slow speed. Open pin 13 gives small bats, and grounding gives large bats.—D. Lancaster, "CMOS Cookbook," Howard W. Sams, Indianapolis, IN, 1977, p 166.

60-LED HYPNOTIC SPIRAL—LEDs are mounted on display board in spiral arrangement and wired in matrix connected to ICs so each LED is lighted in sequence as IC1 and IC2 carry out counting function. IC3 is square-wave oscillator with frequency determined by C1 and setting of R2. Output pulses are used to clock IC1 to advance count, with carry output of IC1 clocking IC2 every tenth count. At end of 60 counts, both ICs reset to zero for new sequence. Inherent current limiting of ICs makes dropping resistors unnecessary for LEDs.—F. Blechman, Digitrance, *Modern Electronics,* Dec. 1978, p 29–31.

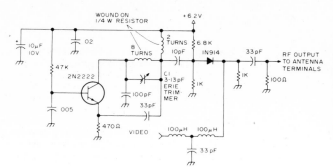

VHF MODULATOR—Developed as interface between General Instruments AY-3-8500-1 TV game chip and antenna terminal of TV set. Adjust C1 to frequency of unused channel to which receiver is set for playing games. Article gives all circuits but covers construction only in general terms.—S. Ciarcia, Hey, Look What My Daddy Built!, *73 Magazine,* Oct. 1976, p 104–108.

OSCILLATOR FOR CHANNELS 2–6—Transmitter serving as interface between video game and TV set can be tuned with L₁ to vacant channel in low TV band. Regular antenna should be disconnected when output of oscillator is fed to TV set via twin-line, to avoid broadcasting game signals. L₁ is 4 turns No. 18 spaced 3/8 inch on ¼-inch slug-tuned form.—B. Matteson, "King Pong" Game Offers Hockey and Tennis Alternatives to TV Re-Runs, *EDN Magazine,* Aug. 5, 1975, p 47–55.

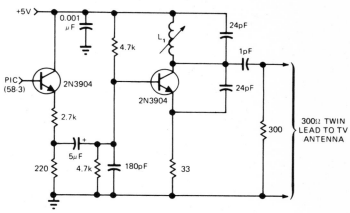

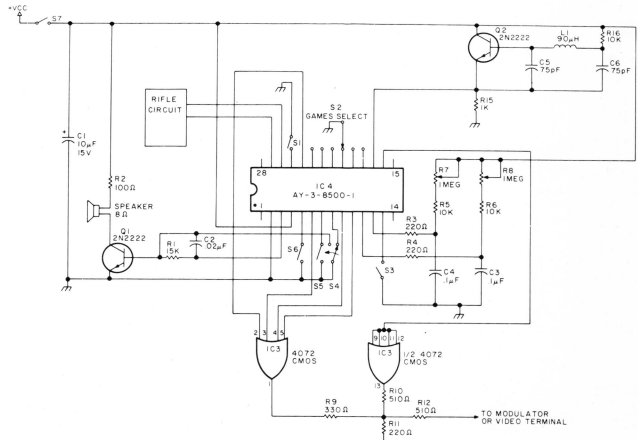

SIX-GAME VIDEO—General Instruments AY-3-8500-1 MOS chip gives choice of hockey, squash, tennis, two types of rifle shoot, and practice games, all with sound effects and automatic scoring on 0-15 display at top of TV screen. Can be used with standard TV receiver (using RF modulator circuit) or with video monitor. S4 grounds base of Q1 when in manual-serve mode, to eliminate steady boing when ball leaves playing field. R5-R8 position players on field. Article covers operation in detail and gives suitable rifle circuit. Supply is +6 V. 2-MHz clock is at upper right.—A. Dorman, Six Games on a Chip, *Kilobaud,* Jan. 1977, p 130, 132, 134, 136, and 138.

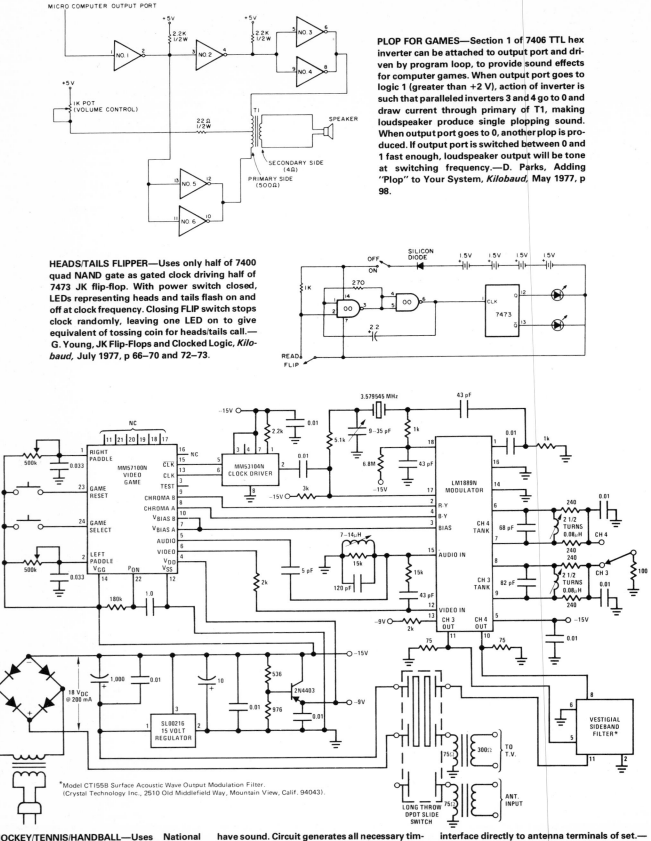

PLOP FOR GAMES—Section 1 of 7406 TTL hex inverter can be attached to output port and driven by program loop, to provide sound effects for computer games. When output port goes to logic 1 (greater than +2 V), action of inverter is such that paralleled inverters 3 and 4 go to 0 and draw current through primary of T1, making loudspeaker produce single plopping sound. When output port goes to 0, another plop is produced. If output port is switched between 0 and 1 fast enough, loudspeaker output will be tone at switching frequency.—D. Parks, Adding "Plop" to Your System, *Kilobaud,* May 1977, p 98.

HEADS/TAILS FLIPPER—Uses only half of 7400 quad NAND gate as gated clock driving half of 7473 JK flip-flop. With power switch closed, LEDs representing heads and tails flash on and off at clock frequency. Closing FLIP switch stops clock randomly, leaving one LED on to give equivalent of tossing coin for heads/tails call.— G. Young, JK Flip-Flops and Clocked Logic, *Kilobaud,* July 1977, p 66—70 and 72—73.

*Model CT155B Surface Acoustic Wave Output Modulation Filter. (Crystal Technology Inc., 2510 Old Middlefield Way, Mountain View, Calif. 94043).

HOCKEY/TENNIS/HANDBALL—Uses National MM57100 TV game chip to provide logic for generating backgrounds, paddles, ball, and digital scoring. All three games are in color and have sound. Circuit generates all necessary timing (sync, blanking, and burst) to interface with circuit of standard TV receiver. With addition of chroma, audio, and RF modulator, circuit will interface directly to antenna terminals of set.— "MOS/LSI Databook," National Semiconductor, Santa Clara, CA, 1977, p 4-37–4-47.

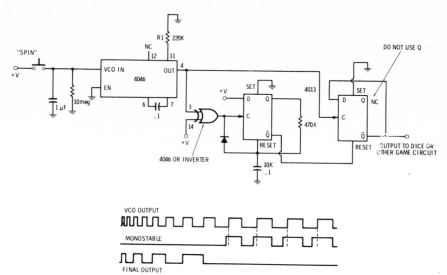

DICE OR ROULETTE RUNDOWN—4046 PLL connected as VCO is set at twice desired maximum rate for dice or roulette-wheel counters. Pressing spin button momentarily to start action charges 1-μF capacitor to supply voltage and jumps VCO to highest frequency. Output frequency then decreases rapidly as capacitor is discharged by 10-megohm resistor. Output is stopped by using retriggerable mono to drive other half of 4013 dual D flip-flop. When frequency drops below value at which mono times out, mono resets flip-flop and holds it to stop display.—D. Lancaster, "CMOS Cookbook," Howard W. Sams, Indianpolis, IN, 1977, p 252–254.

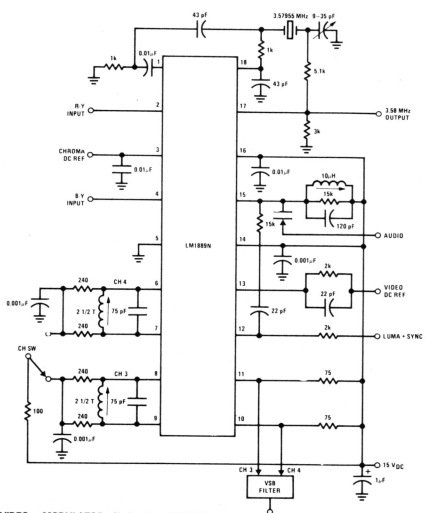

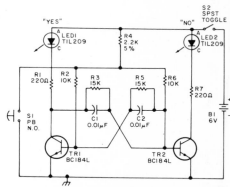

VIDEO MODULATOR—National LM1889N serves to interface audio, color difference, and luminance signals to antenna terminals of TV receiver. Circuit allows video information from video games, test equipment, videotape recorders, and similar sources to be displayed on black-and-white or color TV receivers. LM1889N consists of sound subcarrier oscillator, chroma subcarrier oscillator, quadrature chroma modulators, and RF oscillators and modulators for two low VHF channels.—"MOS/LSI Databook," National Semiconductor, Santa Clara, CA, 1977, p 4-48–4-49.

COIN FLIPPER—One of LEDs comes on when S1 is pressed, to simulate tossing of coin. LEDs can be labeled HEADS and TAILS if desired. Transistor types are not critical. For true random results, voltage between collectors of transistors should be 0 V with S2 closed and S1 open.—Circuits, *73 Magazine*, June 1975, p 161.

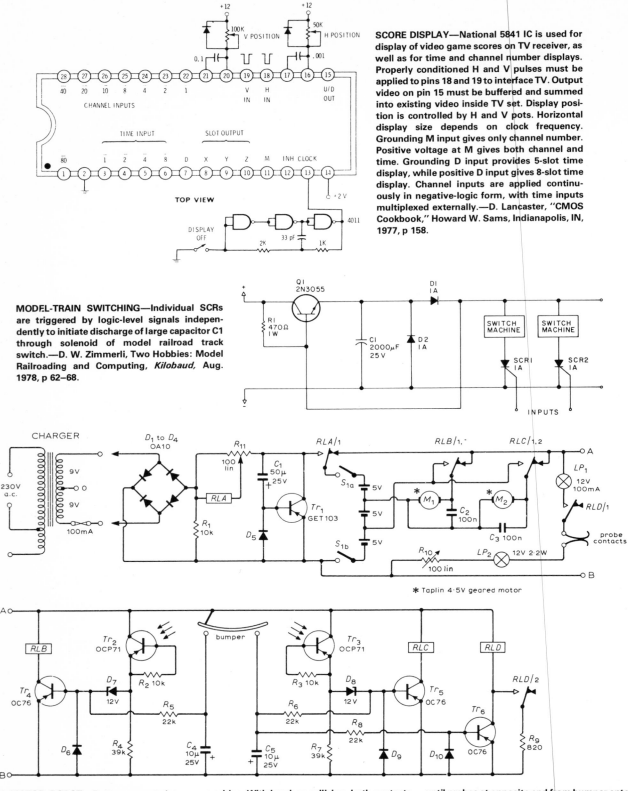

SCORE DISPLAY—National 5841 IC is used for display of video game scores on TV receiver, as well as for time and channel number displays. Properly conditioned H and V pulses must be applied to pins 18 and 19 to interface TV. Output video on pin 15 must be buffered and summed into existing video inside TV set. Display position is controlled by H and V pots. Horizontal display size depends on clock frequency. Grounding M input gives only channel number. Positive voltage at M gives both channel and time. Grounding D input provides 5-slot time display, while positive D input gives 8-slot time display. Channel inputs are applied continuously in negative-logic form, with time inputs multiplexed externally.—D. Lancaster, "CMOS Cookbook," Howard W. Sams, Indianapolis, IN, 1977, p 158.

MODEL-TRAIN SWITCHING—Individual SCRs are triggered by logic-level signals independently to initiate discharge of large capacitor C1 through solenoid of model railroad track switch.—D. W. Zimmerli, Two Hobbies: Model Railroading and Computing, *Kilobaud,* Aug. 1978, p 62–68.

DUAL-MOTOR ROBOT—Battery-operated toy car roams around room, reversing whenever it hits wall or obstacle, and returns automatically to home base when batteries are in need of charge. Small geared motor, such as Meccano No. 11057 or 4.5-V Taplin, is used for each rear wheel so reversal of one motor provides steering. Single free-swiveling caster is at front of machine. With head-on collision, both contacts of bumper close to reverse both motors so machine backs away, turns, and proceeds in new direction. With glancing collision, motor on opposite side is reversed so machine sheers away. White tape on floor, leading to charger having female jacks, is sensed by two phototransistors used to control motors so machine follows tape until probes at opposite end from bumper enter jacks. Circuit permits search mode for recharging only when relay D senses low battery voltage and energizes lamps that illuminate white tape. Article gives operation and construction details.—M. F. Huber, Free Roving Machine, *Wireless World,* Dec. 1972, p 593–594.

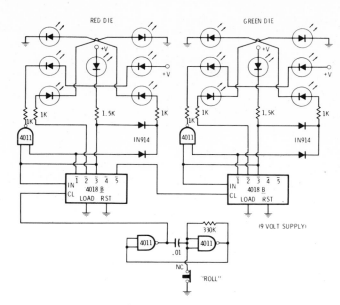

DICE SIMULATOR—Two 4018B synchronous counters are connected in modulo-6 walking-ring sequences for driving LEDs to produce familiar die patterns. Pressing roll button starts gated astable that cycles first die hundreds of times and second die dozens of times, for randomizing of result. When roll button is released, final state of each die is held.—D. Lancaster, "CMOS Cookbook," Howard W. Sams, Indianapolis, IN, 1977, p 324–325.

RANDOM-FLASHING NEONS—Neon glow lamps such as Radio Shack 272-1101 flash in unpredictable sequences at various rates that are determined by values of R and C used for each lamp, to give attention-getting display for classrooms and Science Fairs. Value of R1 can be as low as 2200 ohms for higher repetition rates, but battery drain increases. When circuit is energized, each neon receives full voltage and fires. Lamp capacitor begins charging, decreasing voltage across lamp until lamp goes out and cycle starts over. Use of different capacitor values makes lamps recycle at different rates. T1 is 6.3-VAC filament transformer used to step up oscillator voltage.—F. M. Mims, "Transistor Projects, Vol. 2," Radio Shack, Fort Worth, TX, 1974, p 43–52.

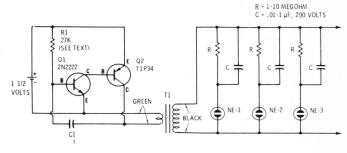

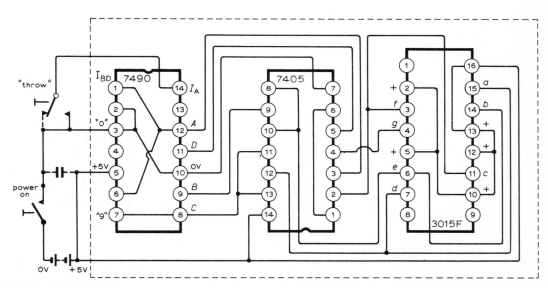

DICE—Simple low-cost arrangement of three ICs operating from 5-V battery (four nickel-cadmium or alkaline cells) provides bar display corresponding to spots on six sides of die. Uses SN7490N TTL decade counter with SN7405 hex inverter to drive Minitron 3015F seven-segment display. Article describes operation in detail and suggests variations for Arabic and binary displays.—G. J. Naaijer, Electronic Dice, *Wireless World,* Aug. 1973, p 401–403.

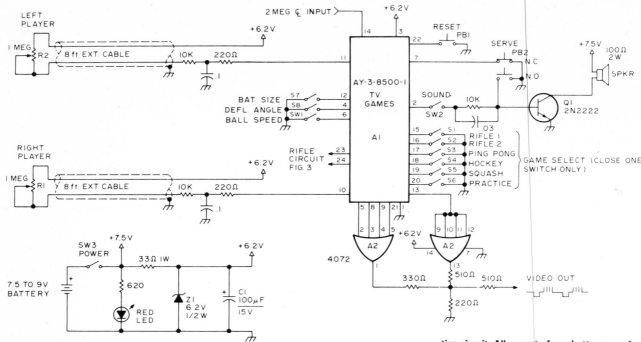

SIX-GAME CHIP—General Instruments AY-3-8500-1 TV game chip and associated circuits give choice of six different games. Article gives additional circuits required, including that for 2- MHz master clock whose output is divided in chip to get vertical and horizontal sync frequencies, VHF modulator used between game and antenna terminal of TV set, and rifle target practice circuit. All operate from battery supply at lower left. Article covers construction only in general terms.—S. Ciarcia, Hey, Look What My Daddy Built!, *73 Magazine*, Oct. 1976, p 104–108.

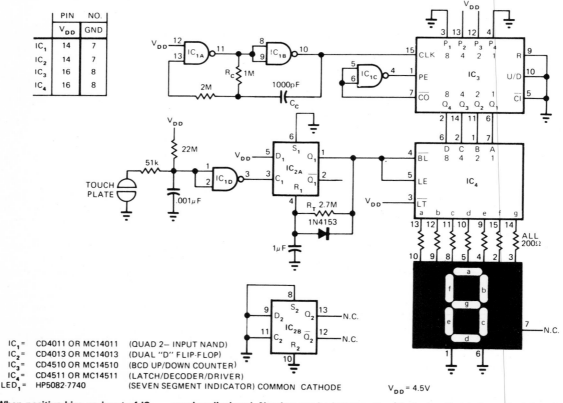

PIN	NO.	
	V_{DD}	GND
IC_1	14	7
IC_2	14	7
IC_3	16	8
IC_4	16	8

IC_1 = CD4011 OR MC14011 (QUAD 2– INPUT NAND)
IC_2 = CD4013 OR MC14013 (DUAL "D" FLIP-FLOP)
IC_3 = CD4510 OR MC14510 (BCD UP/DOWN COUNTER)
IC_4 = CD4511 OR MC14511 (LATCH/DECODER/DRIVER)
LED_1 = HP5082-7740 (SEVEN SEGMENT INDICATOR) COMMON CATHODE

V_{DD} = 4.5 V

LED DIE—When positive bias on input of IC_{1D} NAND gate is pulled to ground by skin resistance of finger, D flip-flop IC_{2A} connected as mono is triggered. Pin 1 goes high for about 2 s, making IC_4 latch outputs of counter IC_3 and unblank LED display. Random time that finger is on touch plate determines randomness of number displayed. Number can be between 1 and 9, between 1 and 6 for die, or between 1 and 2 to represent heads or tails. Change BCD value of jam inputs of IC_3 to highest random number desired. Values shown are for 4.5-V supply and display current of 10 mA per segment. LED is blanked until plate is touched. Standby current drain of 10 μA on three AA alkaline cells is so low that ON/OFF switch is unnecessary.—C. Cullings, Electronic Die Uses Touchplate and 7-Segment LED Display, *EDN Magazine*, May 20, 1975, p 70 and 72.

CHAPTER 8
Intercom Circuits

Covers one-way and two-way basic intercom circuits, four-station two-way system, induction receiver for paging, and private telephone system. Audio amplifier circuits suitable for intercoms are also given.

AUDIO INDUCTION RECEIVER—Used to pick up audio signal being fed to low-impedance single-wire loop encircling room or other area to be covered. Pickup loop L1 is 100–500 turns wound around plastic case of receiver. Opamp sections are from Motorola MC3401P or National LM3900 quad opamp. Supply can be 9–15 V. Requires no FCC license. Can be used as private paging system if audio amplifier of transmitter has microphone input.—C. D. Rakes, "Integrated Circuit Projects," Howard W. Sams, Indianapolis, IN, 1975, p 23–25.

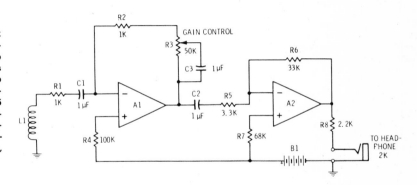

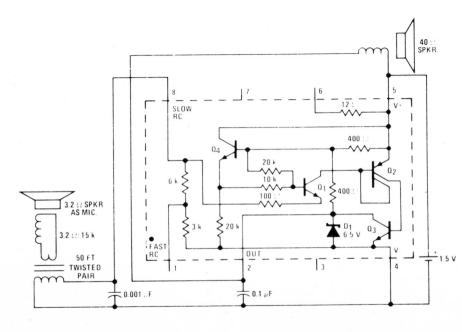

MICROPOWER ONE-WAY INTERCOM—National LM3909 IC operating from single 1.5-V cell serves as low-power one-way intercom suitable for listening-in on child's room and meeting other room-to-room communication needs. Battery drain is only about 15 mA. Person speaking directly into 3.2-ohm loudspeaker used as microphone delivers full 1.4 V P-P signal to 40-ohm loudspeaker at listening location.—"Linear Applications, Vol. 2," National Semiconductor, Santa Clara, CA, 1976, AN-154, p 9.

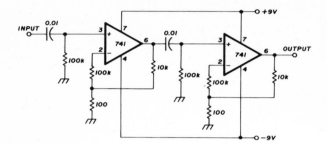

CASCADED 741 OPAMPS—Two opamps in series provide 80 dB of audio gain with bandwidth of about 300 to 6000 Hz. Gain of each opamp is set at 100. With three stages, bandwidth would be 5100 Hz. Output will drive loudspeaker at comfortable room level, if fed through 1-μF non-polarized capacitor to output transformer having 500-ohm primary and 8-ohm secondary.—C. Hall, Circuit Design with the 741 Op Amp, *Ham Radio,* April 1976, p 26–29.

2 W WITH IC—Inexpensive audio amplifier using 14-pin DIP provides adequate power for small audio projects and audio troubleshooting. Pins 3, 4, 5, 10, 11, and 12 are soldered directly to foil side of printed-wiring board used for construction, to give effect of heatsink.—J. Schultz, An Audio Circuit Breadboarder's Delight, *CQ,* Jan. 1978, p 42 and 75.

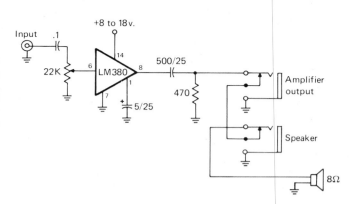

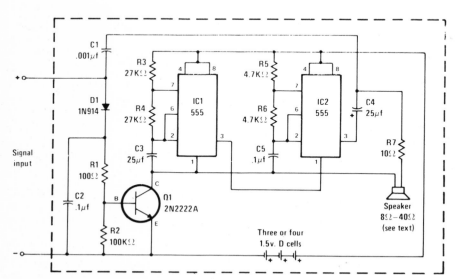

BEEPER—Private two-station telephone system for home requires only two wires between ordinary telephone sets, with 1.5-V battery in series with one line, but this voltage is not enough to actuate ringers in sets. Beeper in parallel with each set, with polarity as shown, serves same purpose as ringer. 555 timer IC1 turns on IC2 about once every 3 s, and IC2 then generates 1000-Hz beep for about 1 s as ringing signal. No switches are required, because telephone handsets provide automatic switching. When both telephones are hung up, 1.5-V battery splits equally between beepers and resulting 0.75 V is not enough to turn on Q1 in either set. When one telephone is picked up, beeper at other telephone receives close to 1.5 V and Q1 turns on IC1 to initiate beeping call. When other telephone is picked up, beeping automatically stops because 1.5 V is again divided between sets.—P. Stark, Private Telephone: Simple Two-Station Intercom, *Modern Electronics,* July 1978, p 32–34.

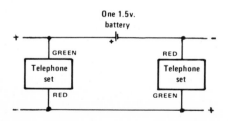

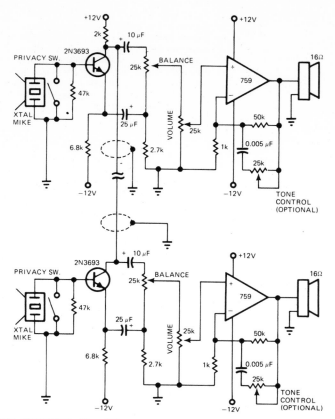

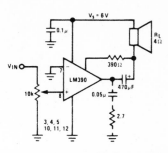

1 W AT 6 V—Battery-operated power amplifier using National LM390 IC provides ample power for loudspeaker despite operation from 6-V portable battery.—"Audio Handbook," National Semiconductor, Santa Clara, CA, 1977, p 4-41.

BIDIRECTIONAL INTERCOM—Uses 759 power opamps to provide 0.5 W for 16-ohm loudspeakers. Crystal microphones feed NPN transistors that provide both in-phase and 180° out-of-phase signals. Balance-adjusting circuits of amplifier cancel out the two signals, so only out-of-phase signal goes to receiving unit. Privacy switch across microphone eliminates audio feedback while listening. Article tells how to calculate heatsink requirements.—R. J. Apfel, Power Op Amps—Their Innovative Circuits and Packaging Provide Designers with More Options, *EDN Magazine*, Sept. 5, 1977, p 141–144.

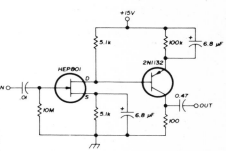

DIRECT-COUPLED AF—Combination of unipolar and bipolar transistors gives desirable amplifying features of each solid-state device. Can be used as speech amplifier and for other low-level audio applications.—I. M. Gottlieb, A New Look at Solid-State Amplifiers, *Ham Radio*, Feb. 1976, p 16–19.

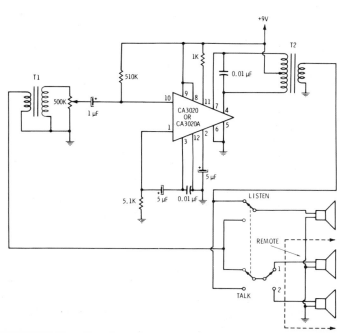

SINGLE IC WITH TRANSFORMERS—CA3020 differential amplifier uses AF input transformer T1 to match loudspeakers (used as microphone) to higher input resistance of IC. AF output transformer T2 similarly matches IC to loudspeakers operating conventionally.—E. M. Noll, "Linear IC Principles, Experiments, and Projects," Howard W. Sams, Indianapolis, IN, 1974, p 100–101.

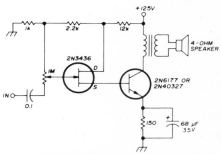

AF OUTPUT—Operates directly from 125-V rectified AC line voltage. Combination of unipolar and bipolar transistors gives desirable amplifying features of each solid-state device.—I. M. Gottlieb, A New Look at Solid-State Amplifiers, *Ham Radio*, Feb. 1976, p 16–19.

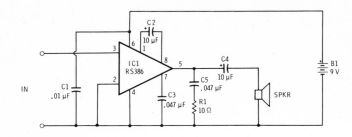

0.25-W AMPLIFIER—Single Radio Shack RS386 IC powered by 6–9 V from battery provides gain of about 200 with sufficient power to drive 8-ohm loudspeaker when speaking closely into small dynamic microphone of type used with portable tape recorders.—F. M. Mims, "Integrated Circuit Projects, Vol. 2," Radio Shack, Fort Worth, TX, 1977, 2nd Ed., p 87–95.

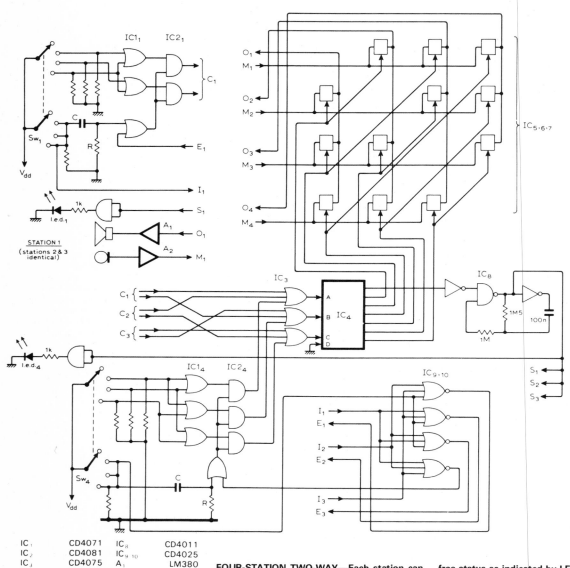

IC$_1$	CD4071	IC$_8$	CD4011
IC$_2$	CD4081	IC$_{9\cdot10}$	CD4025
IC$_3$	CD4075	A$_1$	LM380
IC$_4$	CD4028	A$_2$	741
IC$_{5\cdot6\cdot7}$	CD4016		

Station links	Code
1 to 2	001
1 to 3	010
1 to 4	011
2 to 3	100
2 to 4	101
3 to 4	110

FOUR-STATION TWO-WAY—Each station can communicate privately with any one of others. All four stations have identical inputs as at upper left, with fourth station having master circuit. Each two-station combination is assigned 3-bit code as given in table, for selection by switches Sw$_1$-Sw$_4$. All station codes are ORed and decoded by IC$_4$ to drive matrix of analog switches for coupling appropriate audio inputs and outputs. Code 000 is used for system-free status as indicated by LEDs 1 and 4 being on. LEDs flash for system-busy status. When code is selected, enable inputs of nonselected stations go low to prevent generation of further codes. System can be expanded to six stations by using 4-bit code and CD4514 decoder with larger matrix of analog switches.—B. Voyno-vich, Multiple Station Two-Way Intercom, *Wireless World,* March 1978, p 59.

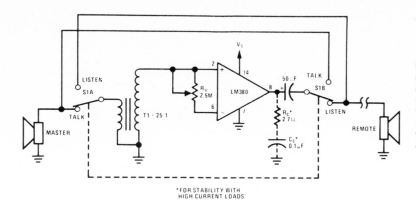

SINGLE OPAMP—When switch S1 is in talk position as shown, loudspeaker of master station acts as microphone, driving opamp through step-up transformer T1. Switch at remote station must then be in listen position. Supply voltage range is 8–20 V.—"Audio Handbook," National Semiconductor, Santa Clara, CA, 1977, p 4-21–4-28.

2-W LM380 POWER AMPLIFIER—Complete basic circuit for most audio or communication purposes uses minimum of external parts. C3, used to limit high frequencies, can be in range of 0.005 to 0.05 μF.—A. MacLean, How Do You Use ICs?, *73 Magazine*, June 1977, p 184–187.

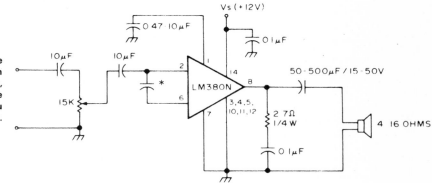

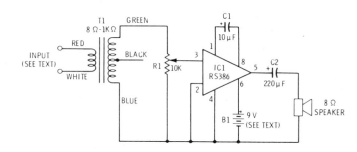

0.5-W AF IC—Simple audio power stage drives 8-ohm loudspeaker for producing greater volume with pocket radio or for intercom applications. Supply range is 4–12 V. For long life, 6-V lantern batteries are recommended. Transformer is Radio Shack 273-1380.—F. M. Mims, "Integrated Circuit Projects, Vol. 5," Radio Shack, Fort Worth, TX, 1977, 2nd Ed., p 38–44.

HIGH-GAIN INTERCOM—Internal bootstrapping in National LM388 audio power amplifier IC gives output power levels above 1 W at supply voltages in range of 6–12 V, with minimum parts count. AC gain is set at about 300 V/V, eliminating need for step-up transformer normally used in intercoms. Optional RC network suppresses spurious oscillations.—"Audio Handbook," National Semiconductor, Santa Clara, CA, 1977, p 4-37–4-41.

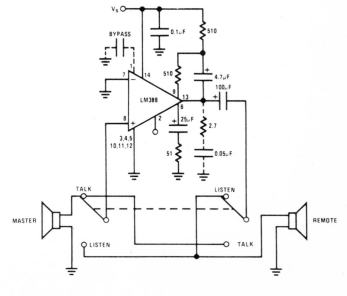

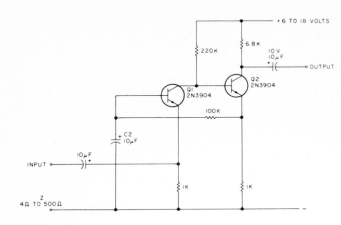

LOW-Z INPUT—Can be used with low-imped-ance source, such as 4- to 16-ohm loudspeaker or telephone earphone used as mike. If loud-speaker is put out in yard, sensitivity is suffi-cient to pick up sounds made by prowlers. Can be fed into input of any high-fidelity amplifier.— E. Dusina, Build a General Purpose Preamp, *73 Magazine,* Nov. 1977, p 98.

CHAPTER 9
Lamp Control Circuits

Covers methods of triggering triacs and silicon controlled rectifiers for turning
on, dimming, and otherwise regulating lamp loads in response to
photoelectric, acoustic, logic, or manual control at input. Starting circuits for
fluorescent lamps are also given.

AC CONTROL WITH TRIAC—Decoder outputs of microprocessor feed 7476 JK flip-flop that drives optocoupler which triggers triac for ON/OFF control of lamp or other AC load. LED and cadmium sulfide photocell are mounted in light shield. When light from LED is on photocell, cell resistance drops and allows control voltage of correct direction and amplitude to trigger gate of triac, turning it on. When light disappears, triac remains on until voltage falls near zero in AC cycle.—R. Wright, Utilize ASCII Control Codes!, *Kilobaud,* Oct. 1977, p 80–83.

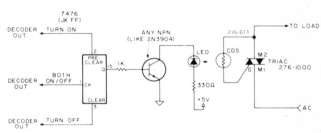

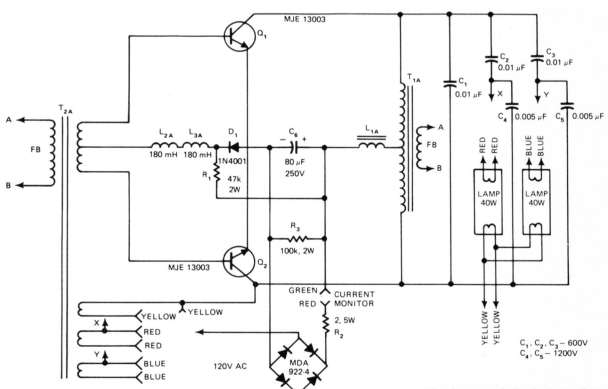

40-W RAPID-START BALLAST—AC line voltage is rectified by diode bridge and filtered by C_6-L_{1A}. Transistors Q_1-Q_2 with center-tapped tank coil T_{1A} and C_1-C_3 make up power stage of 20-kHz oscillator that develops 600 V P sine wave across T_{1A}. When fluorescent lamps ionize, current to each is limited to about 0.4 A. Lamps operate independently, so one stays on when other is removed. Feedback transformer T_{2A} supplies base drive for transistors and filament power for lamps. Article gives transformer and choke winding data.—R. J. Haver, The Verdict Is In: Solid-State Fluorescent Ballasts Are Here, *EDN Magazine,* Nov. 5, 1976, p 65–69.

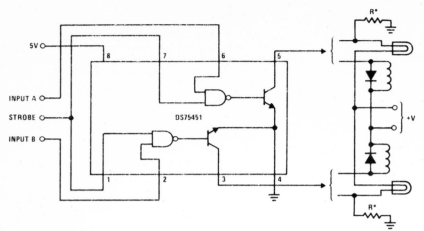

DUAL LAMP DRIVER—National DS75451 dual peripheral AND driver using positive logic provides up to 300 mA per section for driving incandescent lamps. Optional keep-alive resistors R maintain OFF-state lamp current at about 10% of rated value to reduce surge current. Lamp voltage depends on lamps used. Relays shown, with diodes across solenoids, can be used in place of lamps if desired.—"Interface Databook," National Semiconductor, Santa Clara, CA, 1978, p 3-20–3-30.

COMPLEMENTARY FADER—Control unit for stage lighting fades out one lamp while simultaneously increasing light output of another with accurate tracking. Gate of silicon controlled rectifier SCR$_1$ is driven by standard external phase control circuit. Interlock network connected to output of SCR$_1$ provides complementary signal for trigger of SCR$_2$. If lamps larger than 150 W are required, use larger value for C$_1$.—M. E. Anglin, Complementary Lighting Control Uses Few Parts, *Electronics*, Dec. 12, 1974, p 111; reprinted in "Circuits for Electronics Engineers," *Electronics*, 1977, p 78.

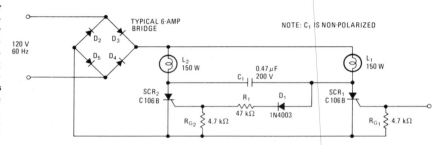

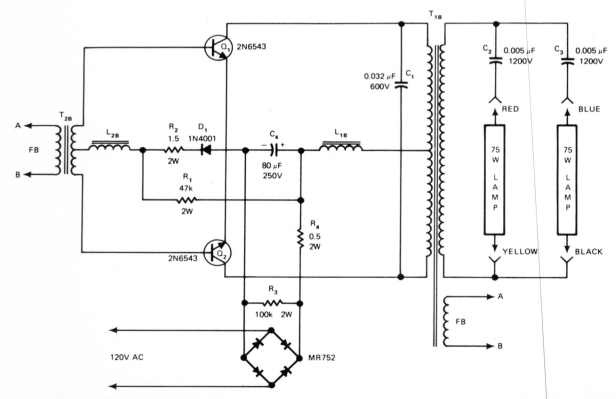

75-W INSTANT-START BALLAST—DC voltage for 20-kHz two-transistor oscillator is obtained from AC line. Secondary is added to center-tapped tank coil of T$_{1B}$ to provide 1 kV P starting voltage required by 96-inch instant-start lamps. Article gives transformer and choke winding data along with circuit details and performance data. Lamps operate independently, so one stays on when other is removed.—R. J. Haver, The Verdict Is In: Solid-State Fluorescent Ballasts Are Here, *EDN Magazine*, Nov. 5, 1976, p 65—69.

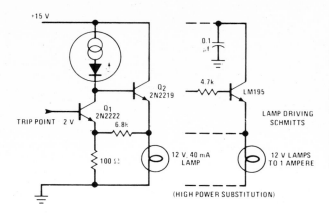

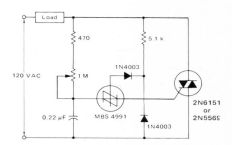

ACTIVE LOAD—National NSL4944 constant-current LED serves as current source for collector resistor of Schmitt trigger to provide up to 12-V output at 40 mA for lamp load. When lamp and Q_2 are off, most of LED current flows through 100-ohm resistor to determine circuit trip point of 2 V. When control signal saturates Q_1, Q_2 provides about 1 V for lamp to give some preheating and reduce starting current surge. When control is above trip point, Q_2 turns on and energizes lamp.—"Linear Applications, Vol. 2," National Semiconductor, Santa Clara, CA, 1976, AN-153, p 3.

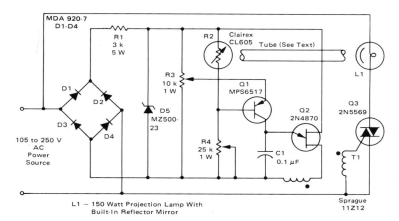

L1 — 150 Watt Projection Lamp With
Built-In Reflector Mirror

800-W TRIAC DIMMER—Simple circuit uses Motorola MBS-4991 silicon bilateral switch to provide phase control of triac. 1-megohm pot varies conduction angle of triac from 0° to about 170°, to give better than 97% of full power to load at maximum setting. Conduction angle is the same for both half-cycles at any given setting of pot.—"Circuit Applications for the Triac," Motorola, Phoenix, AZ, 1971, AN-466, p 5.

PROJECTION-LAMP VOLTAGE REGULATOR—Circuit will regulate RMS output voltage across lamp to 100 V ± 2% for input voltages between 105 and 250 VAC. Light output of 150-W projection lamp is sensed indirectly for use as feedback to firing circuit Q1-Q2 that controls conduction angle of triac Q3. Light pipe, painted black, is used to pick up red glow from back of reflector inside lamp, which has relatively large mass and hence has relatively no 60-Hz modulation.—"Circuit Applications for the Triac," Motorola, Phoenix, AZ, 1971, AN-466, p 12.

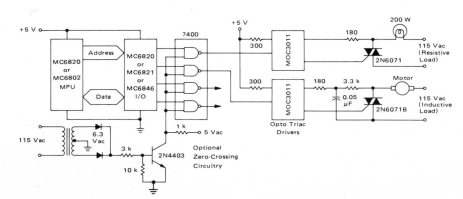

INTERFACE FOR AC LOAD CONTROL—Standard 7400 series gates provide input to Motorola MOC3011 optoisolators for control of triac handling resistive or inductive AC load. Gates are driven by MC6800-type peripheral interface adapters. If second input of two-input gate is tied to simple transistor timing circuit as shown, triac is energized only at zero crossings of AC line voltage. This extends life of incandescent lamps, reduces surge-current effect on triac, and reduces EMI generated by load switching.—P. O'Neil, "Applications of the MOC3011 Triac Driver," Motorola, Phoenix, AZ, 1978, AN-780, p 6.

HELMET-LAMP DIMMER—Provides lossless variation in brightness of incandescent lamp by using duty-cycle modulation. All three sections of 4025 triple three-input NOR gate turn lamp on and off rapidly at rate determined by setting of brightness control pot in astable MVBR circuit. Output transistor rating must be sufficient to handle lamp current.—D. Lancaster, "CMOS Cookbook," Howard W. Sams, Indianapolis, IN, 1977, p 231.

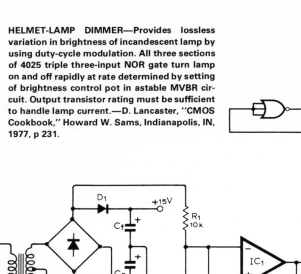

	bass	middle	treble
R21, R22	12k	3k3	820
R25, R26	56k	12k	3k3
R27	1M	220k	56k

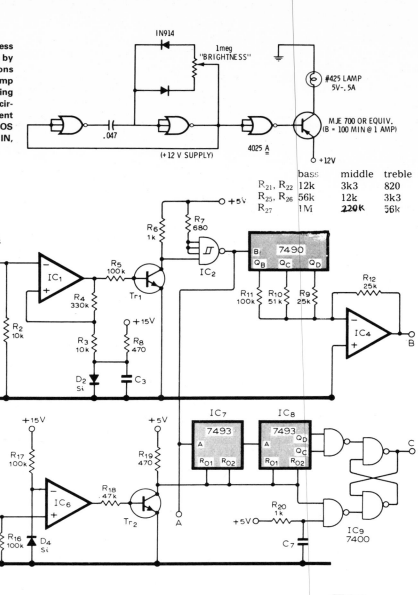

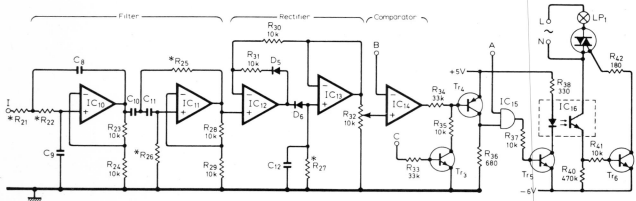

SOUND-CONTROLLED LAMP—Zero-voltage switching achieves interference-free proportional control of lamp intensity by sound source. Both inputs to AND gate IC15 must be high for triac to turn on. One input is from zero-crossing detector IC1, Tr1, and IC2, which produces 100-Hz series of positive-going pulses. Other input is provided by filter/rectifier/comparator circuit. Inverting input of comparator

IC14 is fed by DAC IC4 which produces stepped ramp waveform from outputs of 7490 counter IC3. Counter is connected to count to 5 before resetting internally, giving five possible brightness levels for lamp. Opamps IC5 and IC6 detect when audio input falls below about 10 mV and then release IC7-IC9 from reset stage so the two 4-bit counters start counting 100-Hz waveform. Resetting occurs again when audio input next

passes 10-mV level. Lamp automatically turns on when music stops. All ICs are 741 or equivalent except as marked. Unmarked diodes are 1N4148, C1 and C2 are 100-nF polyester electrolytics, and all transistors are general-purpose types. Resistor values in table are for three-channel system, but more channels can be used if desired.—A. R. Ward, Sound-to Light Unit, *Wireless World,* July 1978, p 75.

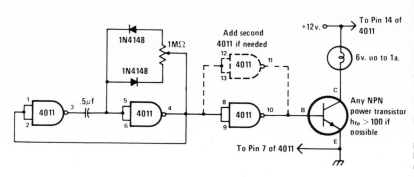

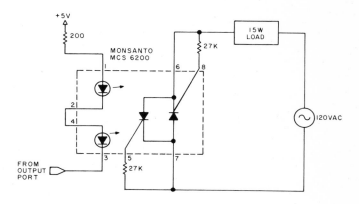

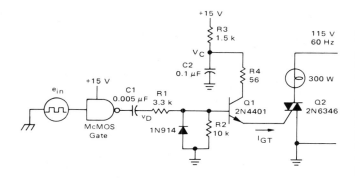

LANTERN-BATTERY EXTENDER—Life of lantern battery can be tripled without reducing light by chopping current while doubling voltage at 50% duty factor. 6-V lamp is connected across chopped 12-V supply built around 4011 CMOS quad NAND gate. First two gates form chopping oscillator, while third serves as interface to any NPN high-gain power transistor. If lamp draws more than 1 A, add fourth gate as shown in dashed lines. If gate is not used, tie its input leads to pin 14 of IC. Duty cycle is varied with 1-megohm pot; set at midrange before applying power, then adjust for normal lamp brilliance.—J. A. Sandler, *9 Easy to Build Projects under $9, Modern Electronics,* July 1978, p 53–56.

FULL-WAVE CONTROL—Monsanto MCS6200 dual SCR optocoupler provides direct full-wave control of 15-W lamp or other AC device when driven by output logic voltages of microprocessor. LEDs are connected in series and photo-SCRs in reverse parallel to create equivalent of triac.—H. Olson, *Controlling the Real World, BYTE,* March 1978, p 174–177.

CMOS LOGIC CONTROL OF 300-W LAMP—Storage capacitor C2 in interface transistor circuit for typical CMOS gate charges to full +15 V supply voltage in time determined by R3 and C2, after which Q1 is fired by positive-going differentiated pulse derived from input square wave. C2 then dumps its charge through R4 and Q1, to fire triac Q2 and energize AC load. For maximum load power, triac should be fired early in conduction angle. With 1-kHz input square wave, output power is over 98% of maximum possible.—A. Pshaenich, "Interface Techniques Between Industrial Logic and Power Devices," Motorola, Phoenix, AZ, 1975, AN-712A, p 13.

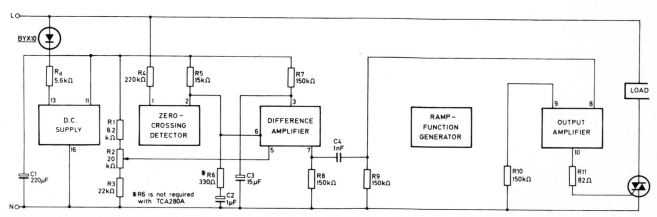

PHASE-CONTROLLED DIMMER—Mullard TCA280A trigger module is connected to compare amplitude of ramp waveform with controllable DC voltage in difference amplifier. At point of coincidence, trigger pulse is produced in output amplifier for triggering triac that controls lamp load. Choice of triac depends on load. Values shown for C4 and R9 give 100-μs pulse.—"TCA280A Trigger IC for Thyristors and Triacs," Mullard, London, 1975, Technical Note 19, TP1490, p 12.

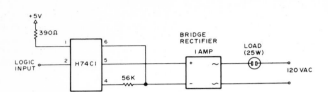

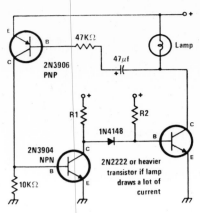

LOGIC CONTROLS 25-W LAMP—Ordinary 1-A bridge is used with H74C1 optoisolator to pass full current to 25-W lamp when logic input goes low (to ground, so full 5 V is applied to light source in optoisolator).—D. D. Mickle, Practical Computer Projects, *73 Magazine,* Jan. 1978, p 92—93.

LAMP SURGE SUPPRESSOR—Circuit limits turn-on current through cold filament, which is major cause of lamp failure but provides normal current when filament reaches operating temperature. Developed primarily for use with lamps in locations where replacement is extremely difficult. Values shown are primarily for low-voltage pilot lamps such as No. 44 and No. 47 but can be applied to any lamp within voltage and current ratings of transistors used.—J. A. Sandler, 11 Projects under $11, *Modern Electronics,* June 1978, p 54—58.

CHAPTER 10
Medical Circuits

Includes circuits for telemetering and processing of heart, brain, muscle, and other bioelectric potentials, recording data from joggers, monitoring therapeutic radiation, synthesizing speech, and providing audible indications for blind persons of light level, voltage, logic status, bridge null, and other measurable parameters.

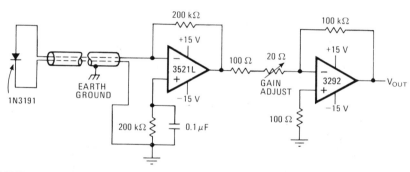

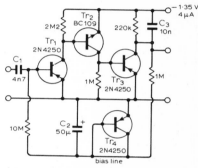

RADIATION MONITOR—1N3191 commercial diode serves as sensor in high-accuracy dosage-rate meter for gamma rays and high-energy X-rays used in radiotherapy. Diode is small enough for accurate mapping of radiation field. Output voltage varies linearly from 0.1 V to 10 V as dose rate increases from 10 to 1000 rads per minute. Low-drift FET-input 3521L opamp amplifies detector current to usable level for 3292 chopper-stabilized opamp that provides additional gain while minimizing temperature errors.—P. Prazak and W. B. Scott, Radiation Monitor Has Linear Output, *Electronics,* March 20, 1975, p 117; reprinted in "Circuits for Electronics Engineers," *Electronics,* 1977, p 106.

IMPLANT AMPLIFIER—Designed for use in implanted transmitters monitoring brain and heart potentials. Requires only 4 μA at 1.35 V. Voltage gain is 2000, and equivalent input noise only 10 μV P-P with 10-megohm source impedance. Tr_1 is current-starved, but resulting limited bandwidth of about 5 kHz is acceptable for biological applications.—C. Horwitz, Micropower Low-Noise Amplifier, *Wireless World,* Dec. 1974, p 504.

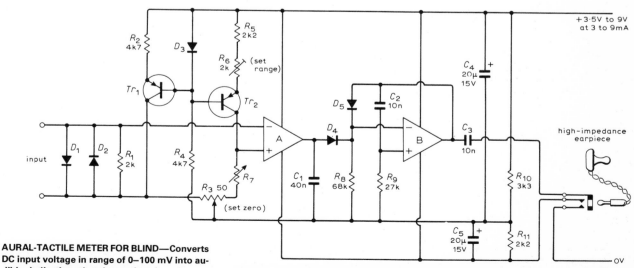

AURAL-TACTILE METER FOR BLIND—Converts DC input voltage in range of 0–100 mV into audible indication that is produced at instant when measured voltage exceeds reference voltage as set by decade switches of R_7. Blind person can then read Braille markings at switch settings to get input voltage. Opamp B is connected as free-running MVBR that generates AF signal for earpiece. Use germanium transistors such as OC45 or OC71. Opamps are Motorola 1435. Use silicon diodes such as 1N914, BA100, or OA200. R_7 can alternatively be wirewound pot.—R. S. Maddever, Meter for Blind Students, *Wireless World,* Jan. 1973, p 36–37.

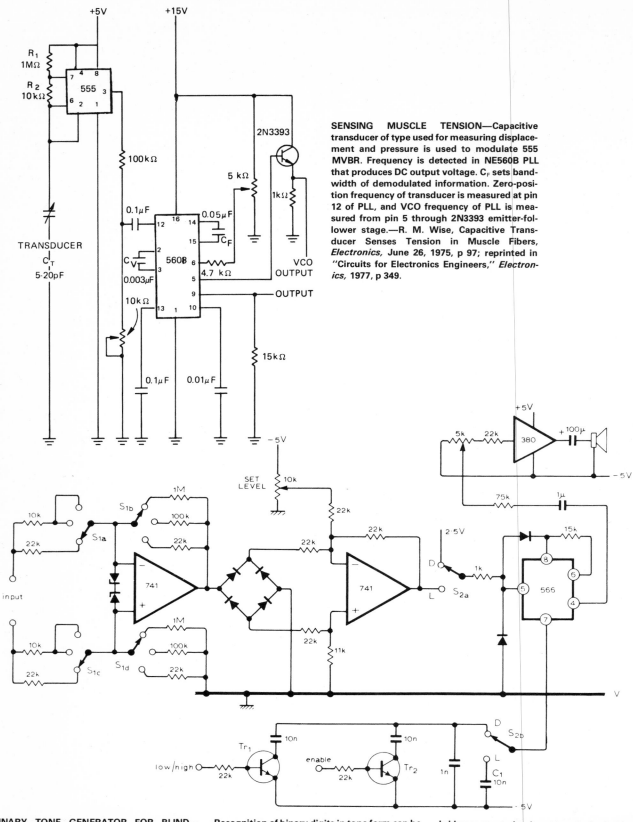

SENSING MUSCLE TENSION—Capacitive transducer of type used for measuring displacement and pressure is used to modulate 555 MVBR. Frequency is detected in NE560B PLL that produces DC output voltage. C_F sets bandwidth of demodulated information. Zero-position frequency of transducer is measured at pin 12 of PLL, and VCO frequency of PLL is measured from pin 5 through 2N3393 emitter-follower stage.—R. M. Wise, Capacitive Transducer Senses Tension in Muscle Fibers, *Electronics*, June 26, 1975, p 97; reprinted in "Circuits for Electronics Engineers," *Electronics*, 1977, p 349.

BINARY TONE GENERATOR FOR BLIND—When low/high input is voltage in binary form, as obtained from converter circuit (also given in article) fed by digital voltmeter, circuit produces low pitch for binary 0 and high pitch for binary 1 when S_2 is set at D for digital voltmeter mode.

Recognition of binary digits in tone form can be learned by blind person much as learning of Morse code. Uses LM566 IC as tone-generating VCO that feeds loudspeaker through LM380 IC amplifier and 5K volume control. With S_2 at position L, circuit serves as audio null detector for

bridge connected to input terminals; S_1 is used to increase sensitivity of 741 opamp as null is approached. Article covers operation of circuits in detail.—R. A. Hoare, An Audible Voltmeter and Bridge-Indicator, *Wireless World*, Sept. 1976, p 87–89.

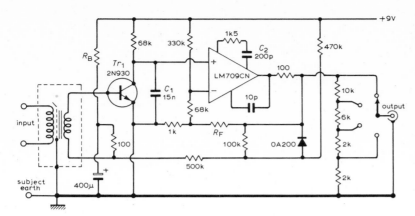

ELECTROMYOGRAM AMPLIFIER—Used to amplify voltages in range of several microvolts to several millivolts in frequency spectrum of 20 to 5000 Hz, as picked up with 13-mm thin silver disks placed on skin over muscle being studied. Article also covers electrocardiographic applications involving source impedances as high as 50 kilohms (as with one electrode on each wrist). Maximum output capability is 9 V P-P. Voltage gain is 1000. R_F is 800K pot, adjusted to give 12 dB per octave dropoff above turnover frequency.—R. E. George, Simple Amplifier for Muscle Voltages, *Wireless World*, Oct. 1972, p 495–496.

AUDIBLE VOLTMETER—Voltage-controlled audio oscillator produces 400-Hz tone for 0 V, with frequency of tone increasing with voltage over two-octave range to 1600 Hz for maximum or full-scale voltage. Ten-resistor voltage divider produces calibrated reference tones corresponding to main 0–10 divisions of meter scale for aural comparison. Simple square-wave audio oscillator Q_1-Q_2 is voltage-controlled by Q_3, which in turn is driven by opamp whose gain is set by R_5. Article covers adjustment of sensitivity pot R_5 and frequency pots R_{24}-R_{33} so VCO tracks voltage being measured and tones coincide at MONITOR and COMPARE positions of S_1 for each meter division.—H. F. Batie, An Audible Meter for the Blind Amateur, *CQ*, Dec. 1973, p 26–31.

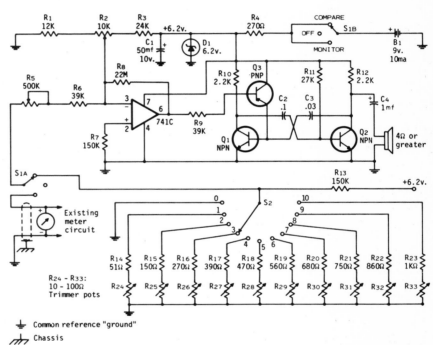

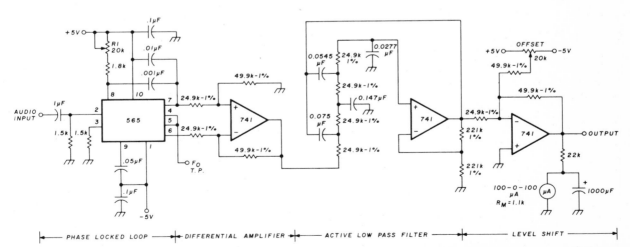

PHASE LOCKED LOOP — DIFFERENTIAL AMPLIFIER — ACTIVE LOW PASS FILTER — LEVEL SHIFT

EKG FM DEMODULATOR—Developed as part of system using satellite for relaying electrocardiograms and other medical data having bandwidth of 0.5 to 50 Hz. Audio signal serving as source of FM is applied to voltage-controlled 1-kHz oscillator having ±40% deviation for full-scale input. Corresponding audio signal at receiving location is fed to input of 565 phase-locked loop. Error voltage of loop, at pin 7, contains data being sought as well as undesirable DC and AC components. DC component of error signal is removed by 741 differential amplifier following PLL. Following four-pole active RC low-pass filter eliminates high-frequency AC components and determines bandwidth of demodulator. Cutoff frequency is 100 Hz. Final 741 opamp scales and shifts output to reasonable value. Recorded output could not be distinguished from original EKG by doctors.—D. Nelson, Medical Data Relay via Oscar Satellite, *Ham Radio*, April 1977, p 67–73.

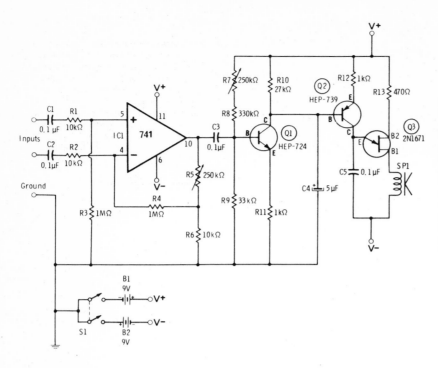

AUDIO EMG MONITOR—Used to measure very small voltages that appear on surface of skin over body muscle. Instead of recording voltage in form of electromyogram (EMG), opamp drives transistor circuit to produce audible note that varies in pitch as EMG signal varies in amplitude. Applications include use by stroke patient as aid to learning reuse of muscle group affected by stroke. Q1 rectifies and averages amplified EMG signal. Q2 controls charging current of C5 for varying frequency of UJT oscillator Q3.—R. Melen and H. Garland, "Understanding IC Operational Amplifiers," Howard W. Sams, Indianapolis, IN, 2nd Ed., 1978, p 125–127.

10-Hz LOW-PASS—Filter design for biomedical experiment has 10-Hz cutoff, tolerable transient and overshoot response, and at least 30-dB rejection of all frequencies above 15 Hz. All components should have 2% tolerance.—D. Lancaster, "Active-Filter Cookbook," Howard W. Sams, Indianapolis, IN, 1975, p 147.

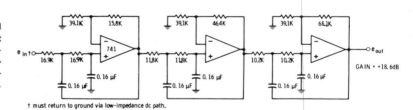

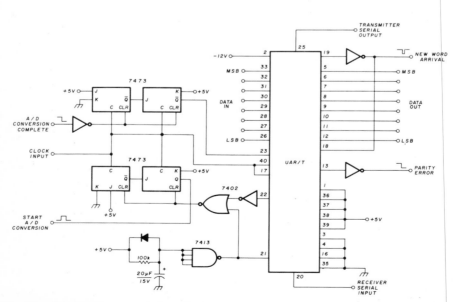

UART FOR EKG RELAY—After electrocardiogram is converted to digital form by commercial A/D converter, circuit shown takes 8-bit word output of converter for processing by universal asynchronous receiver-transmitter (UART) to give required serial asynchronous code for transmitter of satellite relay system, with start, stop, and parity bits added to data under control of 19.2-kHz external clock. This serial output is then used to control FSK oscillator that switches between two discrete audio frequencies to give signal required for transmission through satellite. Article covers operation of UART in detail.—D. Nelson, Medical Data Relay via Oscar Satellite, *Ham Radio,* April 1977, p 67–73.

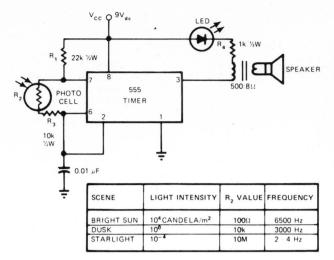

LIGHT-SENSITIVE OSCILLATOR—Uses 555 timer connected so frequency increases directly with intensity of light. Free-running frequency and duty cycle of timer operating in astable mode are controlled by two resistors and one capacitor. R_3 sets upper frequency limit at about 6.5 kHz, and dark resistance of photocell R_2 sets lower limit at about 1 Hz. Loudspeaker provides audio output, while LED flashes for visual indication when frequency goes below about 12 Hz. Applications include detection of lightning flashes, use as optical radar for blind, and use as sunrise alarm.—C. R. Graf, Build a Light-Sensitive Audio Oscillator, *EDN Magazine,* Aug. 5, 1976, p 83.

SCENE	LIGHT INTENSITY	R_2 VALUE	FREQUENCY
BRIGHT SUN	10^4 CANDELA/m^2	100Ω	6500 Hz
DUSK	10^0	10k	3000 Hz
STARLIGHT	10^{-4}	10M	2 - 4 Hz

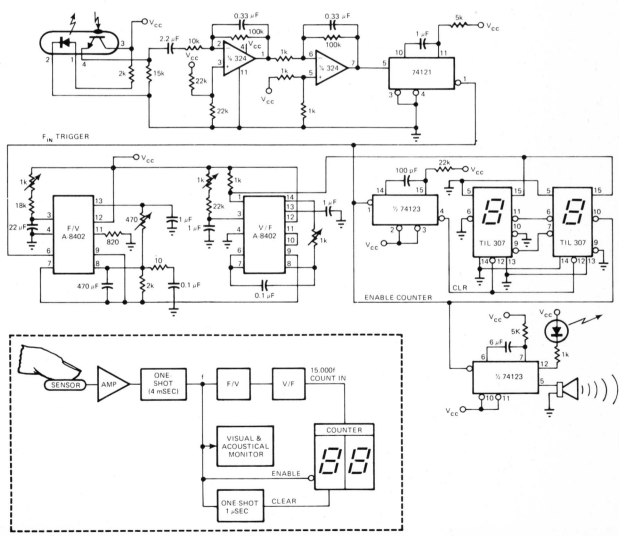

HEART-RATE MONITOR—Measures instantaneous frequency of such slow signals as heart beats (1 Hz) or 33-rpm motors (0.5 Hz) by measuring period T and inverting that quantity to obtain f. Operates from single 5-V supply for portable operation. Fast response time gives reading of heartbeat rate on digital display in two or three pulses. Optoisolator serving as sensor can be taped to almost any part of body because it responds to reflectivity changes caused by changing blood pressure. Accuracy is near 1%.—G. Timmermann, Heartbeat-Rate Monitor Captures VLF Signals, *EDN Magazine,* Oct. 20, 1977, p 79–80.

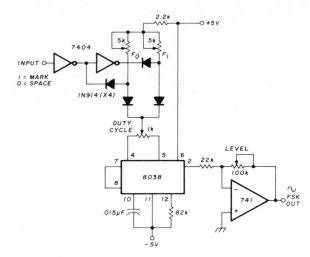

FSK OSCILLATOR FOR EKG RELAY—Used in satellite system for relaying electrocardiograms in digital form. Input consists of 8-bit words obtained in serial form from universal asynchronous receiver-transmitter. Uses 8038 function generator that is switched between two adjustable trimmer resistors giving independently adjustable discrete audio frequencies for mark and space. Output is phase-coherent even though switching does not necessarily take place at zero-crossing points of sine wave. Operation is much like that of FSK RTTY.—D. Nelson, Medical Data Relay via Oscar Satellite, *Ham Radio*, April 1977, p 67–73.

AUDIBLE METER READER—Analog meter terminals are connected to input of DC amplifier Q1 for feeding audio oscillator Q2 and output amplifier Q3. Frequency of oscillator is directly proportional to reading of meter. At calibrate position of S2, DC amplifier is fed by voltage divider R1-R2 and R2 is adjusted until tones heard are identical for both positions of S2. Developed for use by blind person. Knob of R2 sweeps over large scale having markings in Braille for reading of setting at which tones match. Alternatively, R2 can be preset to desired reading and equipment under test adjusted to give tone match. Article covers construction and calibration. C1 is chosen in range of 0.002 to 0.1 μF to give desired minimum fre-

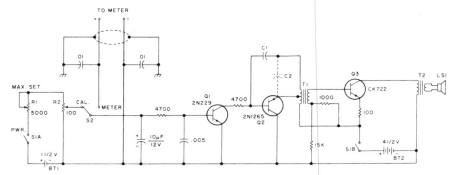

quency. C2, if required, is in same range. T1 is transistor driver transformer (10,000 to 2000 ohms), and T2 is transistor output transformer (500 to 3.2 ohms).—N. Rosenberg, Tune-Up Aids for the Blind, *73 Magazine*, Feb. 1978, p 64–67.

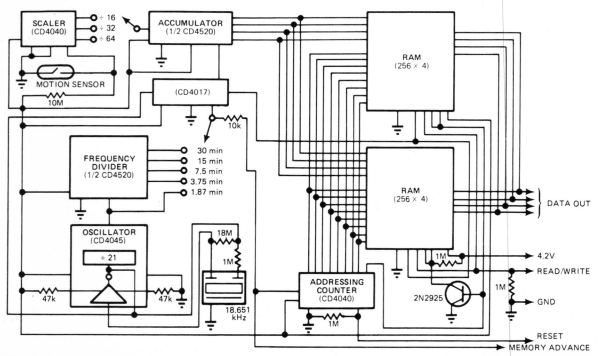

JOGGER DATA COLLECTION—Portable data acquisition system using microcomputer to drive digital cassette tape transport operates from 12-V rechargeable battery and fits in backpack having total weight of only 8 lb. Sample rate can be set between 20 and 100 Hz, with 2 min of continuous data being stored at fast rate. Recorded data is played into PDP-11 minicomputer later for analysis. Motion sensor shown can be replaced by other types of transducers for measuring desired physiological phenomena during jogging, walking, or running.—P. G. Schreier, Physiological Data Acquisition Presents Unusual Problems, Solutions, *EDN Magazine*, June 20, 1978, p 25–26, 28, and 30.

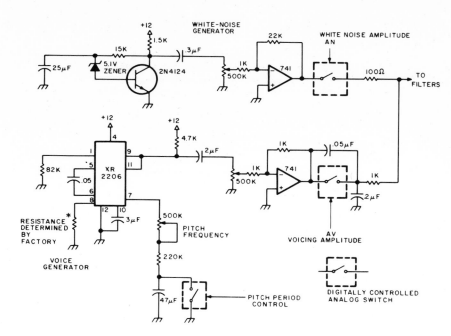

SPEECH SYNTHESIZER—Based on analog simulation of vocal tract. Rush of air through vocal passages is simulated by white-noise generator, while action of larynx is simulated in lower branch of circuit. Article covers problems involved in achieving transitions from phoneme to phoneme, along with automatic emphasis of leading or terminating consonants and intonation of rhythm associated with importance or placement of word in speech. ASCII symbols are given for 33 phonemes generated in Ai Cybernetic Systems model 1000 speech synthesizer, which uses circuit shown in combination with 10 active filters composed of 15 opamps, vocal excitation circuits, ASCII character decoders, and phoneme memories.—W. Atmar, The Time Has Come to Talk, *Byte,* Aug. 1976, p 26–30 and 32–33.

EKG TELEMETER—Developed for experimentation or educational demonstrations in which audience listens to electrocardiograph signal voltage as fed through LM4250 opamp for modulating NE566 connected as VCO driving small loudspeaker. Acoustic output can be picked up by microphone for telemetry purposes if desired. Connection to patient can be made with standard adhesive monitoring electrodes or with small metal disks held on wrists with rubber bands. Tone shifts frequency with each pulse beat.—M. I. Leavey, Inexpensive EKG Encoder, *73 Magazine,* Feb. 1978, p 20–23.

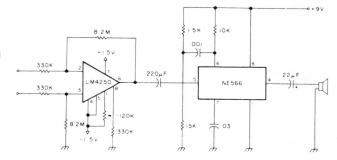

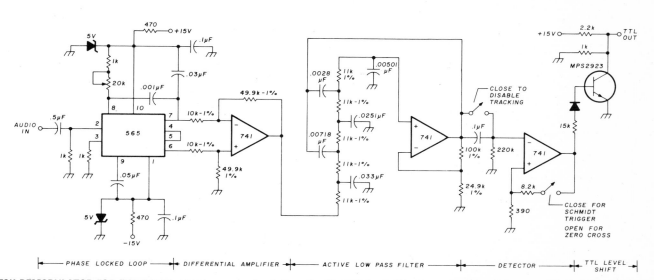

PHASE LOCKED LOOP ← | → DIFFERENTIAL AMPLIFIER ← | → ACTIVE LOW PASS FILTER ← | → DETECTOR ← | → TTL LEVEL SHIFT

FSK DEMODULATOR FOR EKG RELAY—Used at receiving end of satellite system for relaying EKGs, to convert received audio FSK signal to TTL level-shifting output from which original EKG can be obtained. Phase-locked loop tracks input signal frequency and feeds appropriate error signal through differential amplifier to five-pole Butterworth low-pass filter having 1500-Hz cutoff. DC offset is removed by capacitor coupling, for use in zero-crossing detector or Schmitt-trigger detector. Signal is next converted into TTL-compatible level. Recorded output could not be distinguished from original EKG by doctors.—D. Nelson, Medical Data Relay via Oscar Satellite, *Ham Radio,* April 1977, p 67–73.

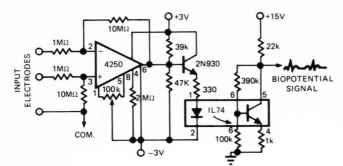

ISOLATED PREAMP—Optoisolator in electro-cardiograph preamp circuit prevents circulating ground currents from shocking patients under test. Can be used with practically all other types of AC line-operated equipment in medical environments.—R. R. Ady, Let's Take an Illuminating Look at Latest Developments in LED's, *EDN Magazine,* Aug. 5, 1975, p 30–35.

AURAL SWR INDICATOR—Permits blind amateur radio operator to check standing waves on transmission line and adjust for best possible impedance match between source and load. Darkened areas are foil strips 6 × 70 mm, 1.5 mm apart, forming inductive trough that transfers RF energy from transmission line to simple aural monitor. Rectified RF energy changes bias on base of Q1, which makes tone increase in pitch with increasing voltage. Idling tone is about 500 Hz for values shown. Operates from three penlight batteries. Transmitter is peaked for maximum output on rising pitch, and matchbox antenna tuner is adjusted for minimum SWR on descending pitch. To lower audio tone, increase size of 82K resistor.—C. G. Bird, Aural SWR Indicator for the Visually Handicapped, *Ham Radio,* May 1976, p 53–53.

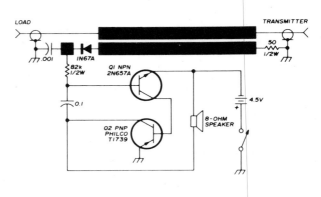

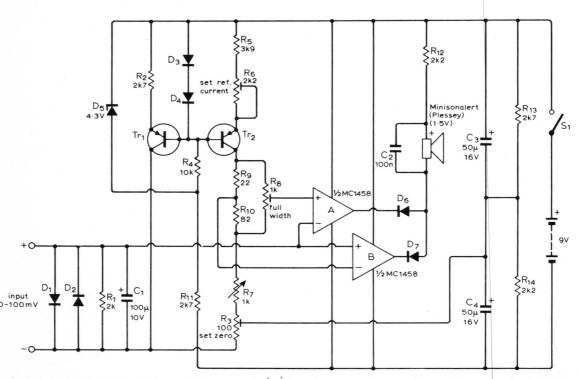

MULTIMETER FOR BLIND—Uses small electric horn to produce sound when DC voltage being measured is different from reference voltage value determined by setting of linear wirewound pot R_7. Blind person adjusts R_7 for null in sound, then reads Braille dots for that setting to get value of voltage being measured. Use PNP silicon transistors, such as BC177 or BC187. D_5 is 4.3-V 400-mW zener, such as BZX79/C4V3. Other diodes are small-signal silicon, such as BA100 or 1N914A.—G. P. Roberts, Multimeters for Blind Students, *Wireless World,* April 1974, p 73–74.

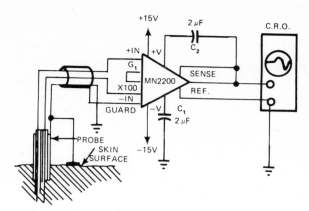

BIOELECTRIC VOLTMETER—Used to measure bioelectric phenomena involving both DC and waveform characteristics with amplitudes of about 10 mV. Since electrodes have impedance of 20,000 to 100,000 ohms, guard terminal must be used to drive input shield. Bias-current return comes from ground plate on skin. Fixed gain of 1000 gives absolute measure of input-voltage magnitude.—R. Duris, Instrumentation Amplifiers—They're Great Problem Solvers When Correctly Applied, *EDN Magazine,* Sept. 5, 1977, p 133–135.

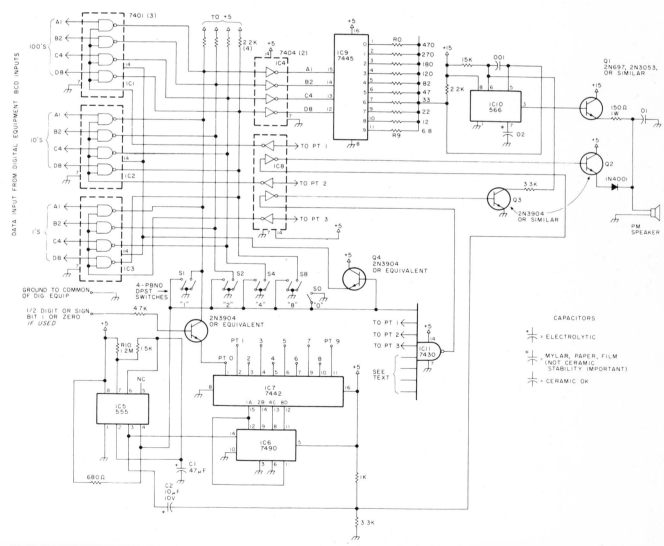

TONE OUTPUT FOR DIGITAL DISPLAY—Converts BCD input from digital test gear to sequence of 10 different tones representing 0 to 9, for recognition of reading on digital display by blind radio operator or experimenter. Length of tone sequence equals number of digits displayed, plus sign indicator or half-digit if desired. Circuit shown is for 3½ digits. Article describes operation of circuit and gives construction details. Resistors R0-R9 (values in kilohms from 6.8K to 470K), determining frequencies of generated tones, are switched into VCO IC10 by IC9.—D. R. Pacholok, Digital to Audio Decoder, *73 Magazine,* Oct. 1977, p 178–180.

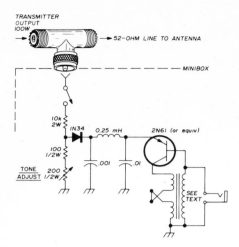

AUDIBLE TUNING FOR BLIND—Transmitter or exciter output is sampled at coax line and high-resistance voltage divider. Rectified voltage of divider, which varies during transmitter tuning, is fed to relaxation oscillator whose output varies in pitch with voltage; low voltage gives high-pitched tone, and high voltage gives low-pitched tone. Input divider draws about 1 W from 100-W transmitter; for higher power, such as 1 kW, change 10K to about 100K. Diode feeds about 2 V to emitter of transistor. Any audio-type PNP transistor can be used. For NPN device, reverse diode connections. Transformer is from 5-W transistor amplifier, with 22-ohm high-impedance winding. Other two windings, in series aiding, are 4 ohms each.—D. H. Atkins, Tuning Aid for the Sightless, *Ham Radio,* Sept. 1976, p 83.

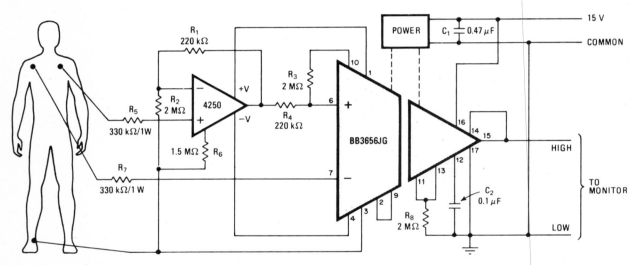

HEART MONITOR—Electrocardiograph amplifier uses Burr-Brown BB3656 isolation amplifier to protect electrocardiograph from inadvertent applications of defibrillation pulses while patient is being monitored. Heart pulses are accurately amplified over frequency range of DC to 3 kHz. Resistors must be carbon-composition types.—B. Olschewski, Unique Transformer Design Shrinks Hybrid Isolation Amplifier's Size and Cost, *Electronics,* July 20, 1978, p 105–11

CHAPTER **11**
Motor Control Circuits

Speed control circuits for various types and sizes of AC and DC motors, including three-phase motors. Some use tachometer feedback to maintain desired constant speed. Includes stepper motor drives, phase sequence detector, braking control, facsimile phase control, and revolution-counting control. Many respond to logic inputs.

SWITCHING-MODE CONTROLLER—Developed for driving 0.01-hp motor M at variable speeds with minimum battery drain. Circuit uses pulses with low duty cycle to set up continuous current in motor approximating almost 200 mA when average battery drain is 100 mA for output voltage of 3.5 V. Voltage comparator A_1 serves as oscillator and as duty-cycle element of controller. C_1 and R_1 provide positive feedback giving oscillation at about 20 kHz, with duty-cycle range of 10% to 70% controlled by feedback loop Q_1-R_1-C_3-R_3. D_2 is used in place of costly large capacitor for filtering.—J. C. Sinnett, Switching-Mode Controller Boosts DC Motor Efficiency, *Electronics*, May 25, 1978, p 132.

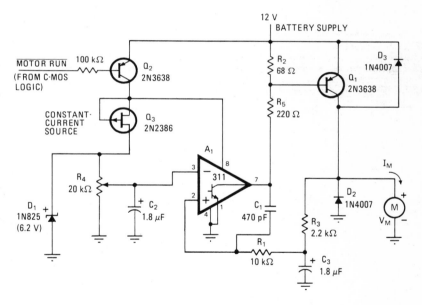

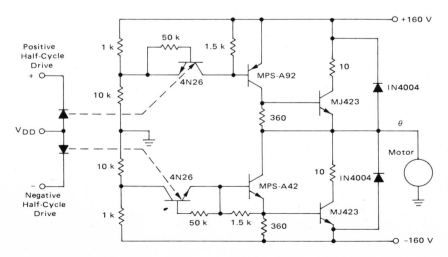

PWM SPEED CONTROL—Power stage using Motorola 4N26 optoisolators and push-pull transistors drives fractional-horsepower single-phase AC motor over speed range of 5% to 100% of base speed. Input drives are provided by pulse-width modulation inverter using stored program in ROM to generate sine-weighted pulse trains to provide variable-frequency drive.—T. Mazur, "A ROM-Digital Approach to PWM-Type Speed Control of AC Motors," Motorola, Phoenix, AZ, 1974, AN-733, p 12.

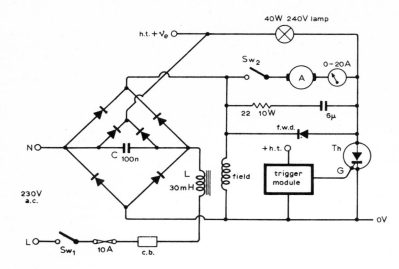

2-hp THYRISTOR CONTROL—Provides smooth variation in speed of shunt-wound DC motor from standstill to 90% of rated speed. Use thyristor rated 30 A at 600 V. Outer diodes of bridge are 35-A 600-PIV silicon power diodes, as also is thyristor diode, and inner diodes are 5-A 600-V silicon power diodes. Article gives complete circuit of trigger pulse generator used to control speed by varying duty cycle of thyristor. Larger motors can be controlled similarly by uprating thyristor and diodes. Controller will also handle other types of loads, including lamps and heaters.—F. Butler, Thyristor Control of Shunt-Wound D.C. Motors, *Wireless World*, Sept. 1974, p 325–328.

TAPE-LOOP SPEED CONTROL—Shunt rectifier-capacitor circuit was developed for speed control of permanent split-capacitor fractional-horsepower induction motor used in some motion-picture projectors. Light-dependent resistor LDR makes Q_2 conduct when light from lamp is not blocked by tape loop. Split capacitor C_1 for motor provides both run and speed-control functions without switching. Values are: C_2 0.01 μF; D 1N4004; Q_1 2N4987; Q_2 C106B; R_1 330K; R_2 100; R_3 10.—T. A. Gross, Control the Speed of Small Induction Motors, *EDN Magazine*, Aug. 20, 1977, p 141–142.

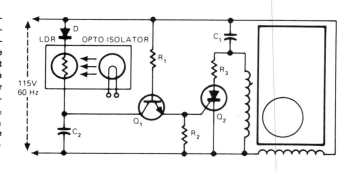

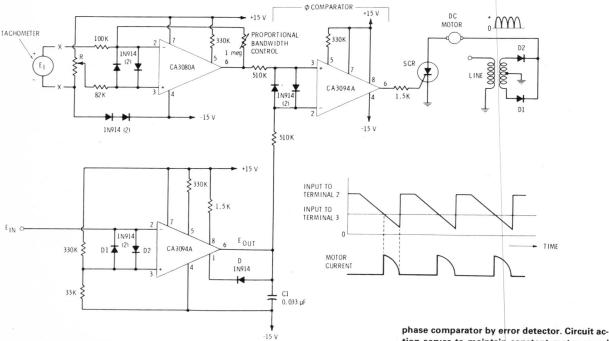

DC MOTOR SPEED CONTROLLER—Tachometer driven by motor produces output voltage proportional to speed for application to CA3080A voltage comparator after rectification and filtering. Output of CA3080A is applied to upper CA3094A phase comparator that is receiving reference voltage from another CA3094A connected as ramp generator. Output of phase comparator triggers SCR in motor circuit. Amount of motor current is set by time duration of positive signal at pin 6, which in turn is determined by DC voltage applied to pin 3 of phase comparator by error detector. Circuit action serves to maintain constant motor speed at value determined by position of pot R. Input to ramp generator is pulsating DC voltage used to control rapid charging of C1 and slower discharging to form ramp.—E. M. Noll, "Linear IC Principles, Experiments, and Projects," Howard W. Sams, Indianapolis, IN, 1974, p 321–323.

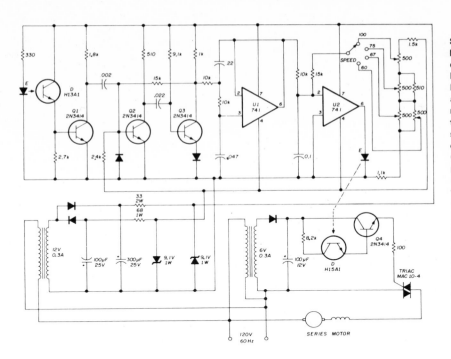

SERIES-MOTOR SPEED CONTROL—Adjustable-speed solid-state motor drive replaces governor in Kleinschmidt RTTY page printer, to give knob-controlled speed range of 60 to 100 WPM. Notched (33-slot) sheet-aluminum disk serving as pulse wheel is mounted on motor shaft and rotates in gap between LED and phototransistor of GE H13A1 optical coupler to form motor-speed sensor or tachometer. Pulses from tachometer, squared by Q1, trigger mono MVBR Q2-Q3 which converts signal to constant-amplitude constant-width pulses having repetition rate proportional to motor speed. Opamp U1 forms three-pole Butterworth active filter that develops required average DC voltage from pulse train. Output current of U1 is compared to reference current derived from speed control circuit, for switching U2 sharply on and off as speed varies above and below desired value. U2 in turn switches motor on and off through H15A1 optical coupler and Q4 in gate circuit of triac. Second coupler isolates control circuit from AC line.—K. H. Sueker, Electronic Speed Control for RTTY Machines, *Ham Radio*, Aug. 1974, p 50–54.

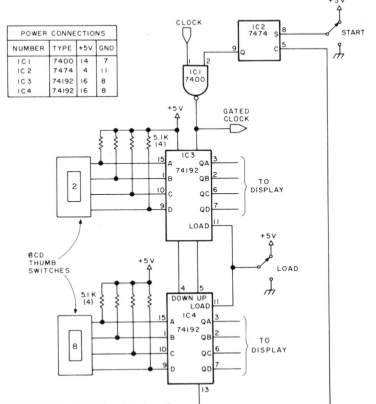

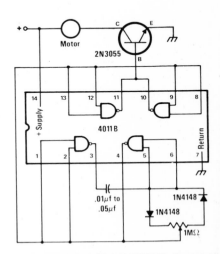

POWER CONNECTIONS			
NUMBER	TYPE	+5V	GND
IC 1	7400	14	7
IC 2	7474	4	11
IC 3	74192	16	8
IC 4	74192	16	8

STEP COUNTER FOR STEPPER—Used to deliver selected number of pulses to stepper motor when start button is pushed, in microprocessor application where number of steps is more important than precise speed. Thumbwheel switch inputs can be I/O port lines of microprocessor. LOAD line transfers into counter the desired count as set up by switches. Article gives flowcharts and software routines for microprocessor to be used for controlling stepper motor.—R. E. Bober, Taking the First Step, *BYTE*, Feb. 1978, p 35–36, 38, 102, 104, 106, and 108–112.

SPEED CONTROL FOR 3-V MOTOR—Designed for use with hobby or toy motors running at about 10,000 rpm and powered by 3-V to 6-V batteries. Uses 4011 CMOS NAND gate with diodes and power transistor to provide variable duty cycle, so adjustment of 1-megohm speed control varies average voltage applied to motor without affecting peak voltage. Motor battery is connected between + terminal and ground of circuit.—J. A. Sandler, 11 Projects under $11, *Modern Electronics*, June 1978, p 54–58.

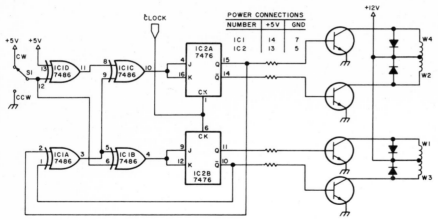

FOUR-PHASE STEPPER DRIVE—EXCLUSIVE-OR gates of 7486 provide steering, while 7476 flip-flops provide memory for generating drive patterns of bidirectional logic stepper motor that is controlled by microprocessor. Output transistors, diodes, and resistors are chosen to meet power requirements for each phase of motor. Speed is controlled by frequency of clock input. Use 555 for coarse control or crystal oscillator for accurate control. S1, which can be an I/O line of microprocessor, controls direction of rotation. Frequency can be obtained from digitally controlled oscillator whose setting is determined by DAC.—R. E. Bober, Taking the First Step, *BYTE*, Feb. 1978, p 35–36, 38, 102, 104, 106, and 108–112.

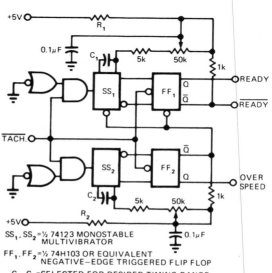

UNDER/OVERSPEED LOGIC—Provides signal (READY output high) only when tachometer pulses from motor are within specific upper and lower limits. Also provides overspeed output signal when upper limit has been exceeded. Single-action triggering eliminates instability at decision point. Article covers circuit operation in detail and gives timing diagram.—W. Bleher, Circuit Indicates Logic "Ready," *EDN Magazine*, March 5, 1974, p 72 and 74.

SS_1, SS_2 = ½ 74123 MONOSTABLE MULTIVIBRATOR

FF_1, FF_2 = ½ 74H103 OR EQUIVALENT NEGATIVE–EDGE TRIGGERED FLIP FLOP

C_1, C_2 = SELECTED FOR DESIRED TIMING RANGE

R_1 = SELECTED FOR UNDER/NOMINAL SPEED HYSTERESIS

R_2 = SELECTED FOR NOMINAL/OVER SPEED HYSTERESIS

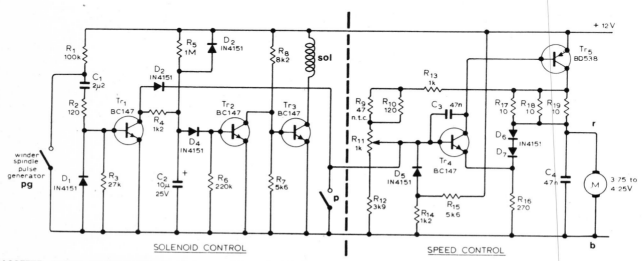

SOLENOID CONTROL   SPEED CONTROL

CASSETTE DRIVE CONTROLLER—Used in high-quality stereo cassette deck operating from AC line or battery. Combines current source for cassette-retaining solenoid with speed control for drive motor. As motor turns, associated motor-driven pulse-generating switch keeps Tr_1 conducting; this cuts off Tr_2 and allows current to flow through Tr_3 for en-ergizing solenoid. When motor stops, pulse-generating switch also stops and Tr_1 stops conducting. After 3-s delay determined by C_2 and R_5, Tr_2 conducts and solenoid is deenergized, releasing cassette. In speed control circuit, Tr_5 acts as constant-current source for motor, using feedback from its collector to base of Tr_4. Back EMF developed by motor is applied to emitter of Tr_4, reducing its forward bias and reducing current in the base of Tr_5 so as to stabilize motor. Article gives all other circuits of cassette deck and describes operation in detail.—J. L. Linsley Hood, Low-Noise, Low-Cost Cassette Deck, *Wireless World*, Part 3—Aug. 1976, p 55–56 (Part 1—May 1976, p 36–40; Part 2—June 1976, p 62–66).

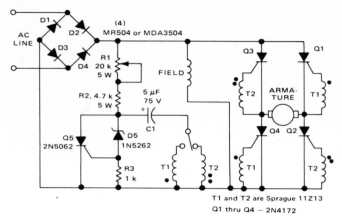

SHUNT-WOUND MOTOR—Switch provides direction control and R1 controls speed of fractional-horsepower shunt-wound DC motor. Field is placed across rectified supply, and armature windings are in four-SCR bridge circuit. Switch determines which diagonal pair of SCRs is turned on, to control direction of rotation. Triggering circuit consisting of Q5, D5, and C1 is controlled by R1, for changing conduction angle of triggered SCR path.—"Direction and Speed Control for Series, Universal and Shunt Motors," Motorola, Phoenix, AZ, 1976, AN-443.

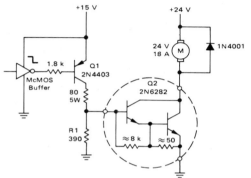

LOW-LEVEL CMOS CONTROL—Low-level output of CMOS buffer turns on DC motor through Q1 and 20-A Darlington power transistor Q2.— A. Pshaenich, "Interface Techniques Between Industrial Logic and Power Devices," Motorola, Phoenix, AZ, 1975, AN-712A, p 18.

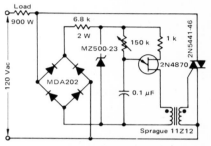

900-W FULL-WAVE TRIGGER—Uses UJT for phase control of triac. Suitable for control of shaded-pole motors driving loads having low starting torque, such as fans and blowers.—D. A. Zinder, "Electronic Speed Control for Appliance Motors," Motorola, Phoenix, AZ, 1975, AN-482, p 4.

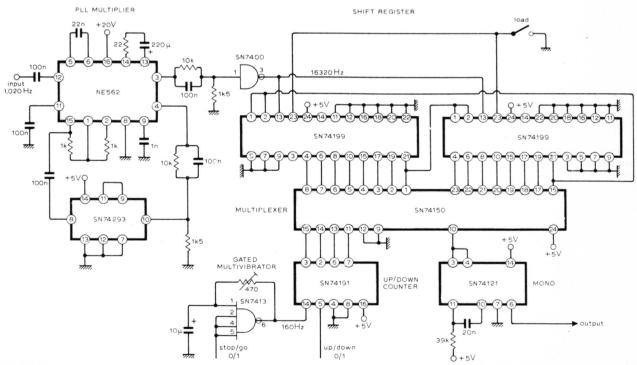

FACSIMILE PHASE CONTROL—Circuit provides accurate phasing of 51-pole-pair phonic/synchronous motor in facsimile transmitter, and can readily be adapted for similar applications. A 16-stage shift register loaded with 1 bit and connected as ring counter is clocked at 16 times required drive motor frequency. This gives pulse train with 1:15 mark-space ratio and repetition rate equal to drive frequency. Multiplexer used as single-pole 16-way switch can select output for any stage of shift register; each clockwise switch step gives 360/16 or 22.5° phase advance. Article describes circuit operation in detail.—P. E. Baylis and R. J. Brush, Synchronous-Motor Phase Control, *Wireless World*, April 1976, p 62.

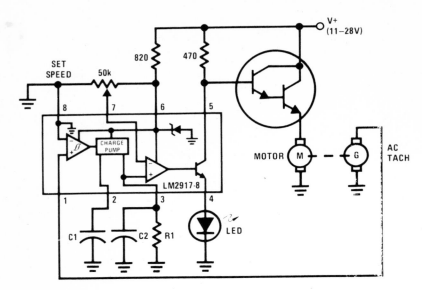

SHUNT-MODE SPEED CONTROL—AC tachometer on shaft of DC motor serves as input for National LM2917N-8 IC acting as shunt-mode regulator with LED indicator. Output of Darlington power transistor provides analog drive to motor. As motor speed approaches reference level set by values chosen for R1, C1, and C2, current to motor is proportionately reduced so motor comes gradually up to speed and is maintained there without operating in switching mode. Advantage of this arrangement is absence of electric noise normally generated during switching-mode operation.—"Linear Applications, Vol. 2," National Semiconductor, Santa Clara, CA, 1976, AN-162, p 10–11.

PAPER-TAPE FEED—High or 1 bit at output port of microprocessor turns on LED of optocoupler to energize solenoid of pinch-roller drive for paper tape of tape reader. Circuit will control reader from computer keyboard. Optoisolator is essential to keep grounds separate, since mechanical devices are electrically noisy and can generate garbage in computer. Article gives software for tape input routine on 8008 microprocessor.—D. Hogg, The Paper Taper Caper, *Kilobaud,* March 1977, p 34–40.

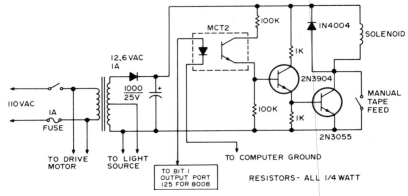

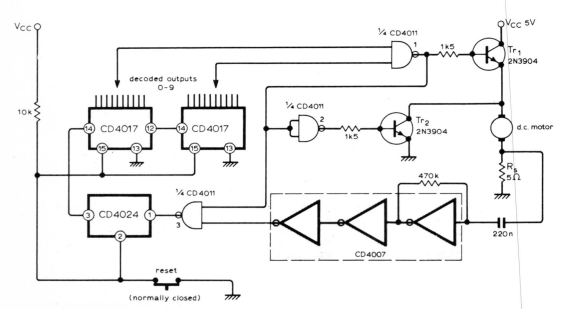

REVOLUTION-COUNTING CONTROL—When desired number of revolutions is reached by DC motor, as determined by preset counter, Tr_1 is turned off to interrupt path to 5-V motor supply, while TR_2 is turned on to brake motor rapidly. Voltage developed across 5-ohm resistor R_s in series with motor contains frequency component related to speed of rotation and number of armature coils. This signal is amplified by CD4007 CMOS inverter for feeding to counters through signal-squaring inverters. Counter outputs are decoded by gate 1. Motor slowdown by heavy loads does not affect accuracy of revolution-counting.—R. McGillivray, Motor Revolutions Control, *Wireless World*, Jan. 1977, p 76.

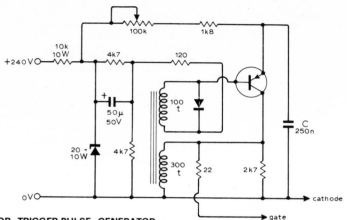

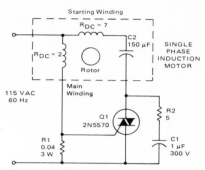

THYRISTOR TRIGGER-PULSE GENERATOR— Used with thyristor speed control for 2-hp shunt-wound DC motor. Circuit provides train of pulses with variable delay with respect to zero-crossing instants of AC supply, for feeding to cathode and gate of thyristor to vary duty cycle. Use Mullard BFX29 silicon PNP transistor or equivalent, and any small-signal silicon diode. Output pulses are suitable for triggering all types of thyristors up to largest. Article also gives motor control circuit.—F. Butler, Thyristor Control of Shunt-Wound D.C. Motors, *Wireless World,* Sept. 1974, p 325–328.

TRIAC STARTING SWITCH FOR ½-hp MOTOR— Triac replaces centrifugal switch normally used to control current through starting winding of single-phase induction motor. Value of R1 is chosen so triac turns on only when starting current exceeds 12 A. When motor approaches normal speed, running current drops to 8 A and triac blocks current through starting winding.— "Circuit Applications for the Triac," Motorola, Phoenix, AZ, 1971, AN-466, p 8.

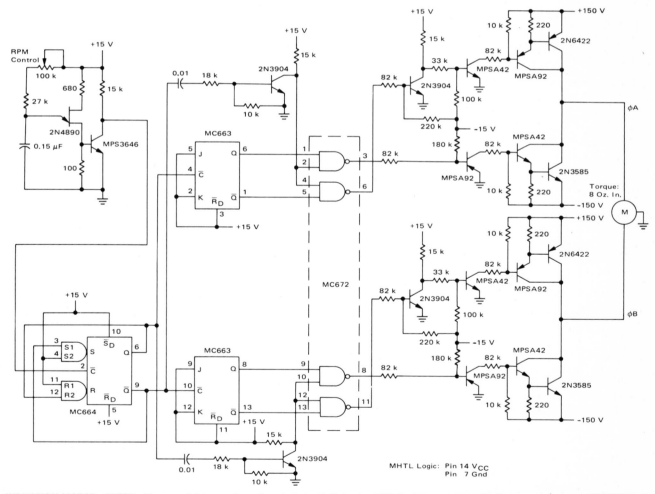

INDUCTION-MOTOR SPEED—Uses variable-frequency UJT oscillator at upper left to toggle MC664 RS flip-flop which in turn clocks MC663 JK flip-flops. Quadrature-phased JK outputs are combined with fixed-width pulses in MC672 to provide zero-voltage steps of drive signals for phase A and phase B. Outputs of RS flip-flops are differentiated and positive-going transitions amplified by pair of 2N3904 transistors, with pulse width of about 500 μs. NAND-gate outputs are then translated by small-signal amplifiers to levels suitable for driving final transistors having complementary NPN/PNP pairs. Circuit will provide speed range of 300 to 1700 rpm for permanent-split capacitor motor.—T. Mazur, "Variable Speed Control System for Induction Motors," Motorola, Phoenix, AZ, 1974, AN-575A, p 6.

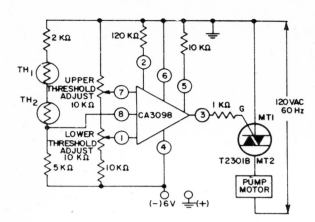

WATER-LEVEL CONTROL—Two thermistors operating in self-heating mode are mounted on sides of water tank. Thermistors change resistance when water level rises so liquid rather than air conducts heat away. Threshold adjustment pots are set so RCA CA3098 programmable Schmitt trigger turns on pump motor when water level rises above thermistor mounted near upper edge of tank, to remove water from tank and prevent overflow. Motor stays on to pump water out of tank until water level drops below location of lower thermistor inside tank.—"Linear Integrated Circuit and MOS/FET's," RCA Solid State Division, Somerville, NJ, 1977, p 218–221.

OPAMP SPEED CONTROL—Provides fine speed control of DC motor by using 0.25-W 6-V motor as tachogenerator giving about 4 V at 13,000 rpm. Opamp (RCA 3047A or equivalent) provides switching action for transistor in series with controlled motor, up to within a few volts of supply voltage. Choose transistor to meet motor current requirement.—N. G. Boreham, D.C. Motor Controller, *Wireless World,* Aug. 1971, p 386.

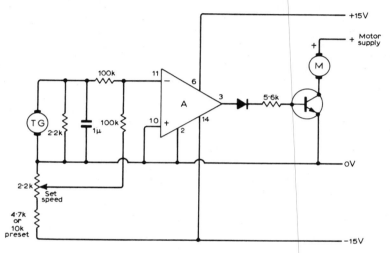

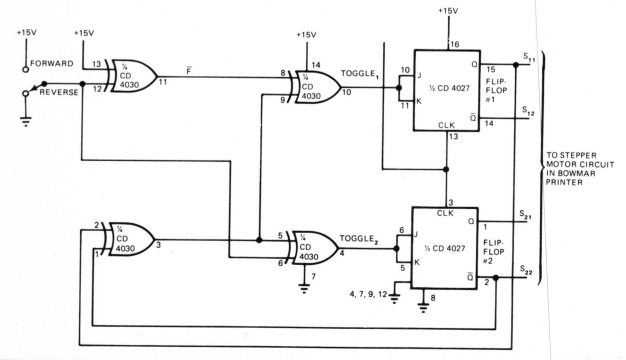

STEPPER MOTOR DRIVE—Two CMOS packages provide the four feed signals required for controlling forward/reverse drive of stepper motor for carriage drive and paper advance of Bowmar Model TP 3100 thermal printer. Outputs of flip-flops are above 10 V, enough to drive stepper motor directly. Each clock pulse to JK flip-flop advances carriage one step in direction commanded.—R. Bober, Stepper Drive Circuit Simplifies Printer Control, *EDN Magazine,* April 5, 1976, p 114.

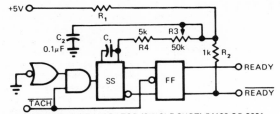

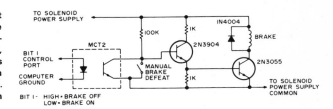

UP-TO-SPEED LOGIC—Simple speed-sensing circuit fed by tachometer pulses makes READY output high when rotating device reaches desired minimum or threshold speed. Single-action triggering eliminates instability at decision point. Circuit also provides hysteresis, for separating pull-in and drop-out points any desired amount as determined by ratio of R_1 to R_2 in timing network. Article covers circuit operation and gives timing diagram.—W. Bleher, Circuit Indicates Logic "Ready," *EDN Magazine,* March 5, 1974, p 72 and 74.

SS=MONOSTABLE MULTIVIBRATOR (SINGLE SHOT) 74122 OR 9601 OR ½ 74123 OR 9602
FF=NEGATIVE EDGE–TRIGGERED FLIP FLOP, 74H103 OR EQUIVALENT
R_1=0 TO 20% OF R_2, SELECTED FOR REQUIRED HYSTERESIS
C_1=SELECTED FOR REQUIRED TIMING RANGE.

CONTINUOUS-DUTY BRAKE—High or 1 bit at output port of microprocessor energizes brake solenoid of paper-tape reader through optocoupler and amplifier. When tape is to be stopped, brake solenoid is energized and tape is squeezed between top of solenoid and flat iron brake shoe that is attracted by solenoid.—D. Hogg, The Paper Taper Caper, *Kilobaud,* March 1977, p 34–40.

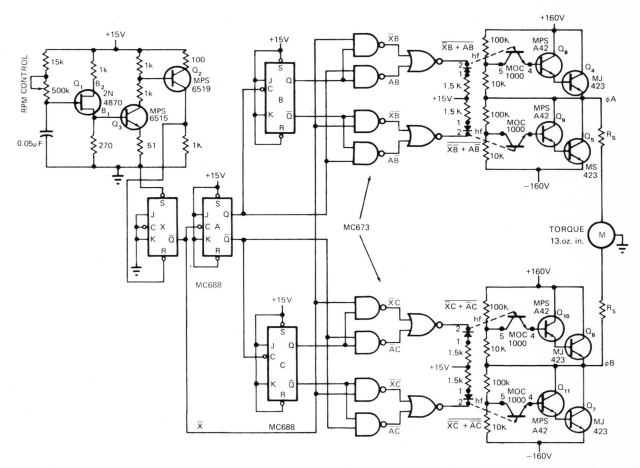

SPEED CONTROL FOR INDUCTION MOTOR— Uses UJT oscillator Q_1 to generate frequency in range from 40 to 1200 Hz for feeding to divide-by-4 configuration that gives motor source frequency range of 10 to 300 Hz. With induction motor having two pairs of poles, this gives theoretical speed range of 300 to 9000 rpm with essentially constant torque. Speed varies linearly with frequency. Circuit uses pair of flip-flops (MC673) operated in time-quadrature to perform same function as phase-shifting capacitor so motor receives two drive signals 90° apart. Article covers operation of circuit in detail. Optoisolators are used to provide bipolar drive signals from unipolar control signals. Each output drive circuit is normally off and is turned on only when its LED is on. If logic power fails, drives are disabled and motor is turned off as fail-safe feature.—T. Mazur, Unique Semiconductor Mix Controls Induction Motor Speed, *EDN Magazine,* Nov. 1, 1972, p 28–31.

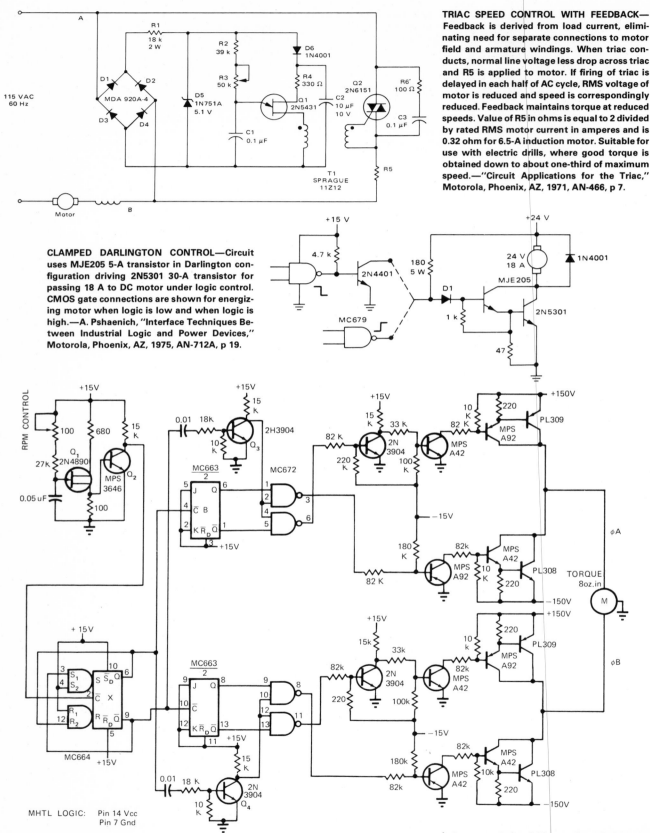

TRIAC SPEED CONTROL WITH FEEDBACK— Feedback is derived from load current, eliminating need for separate connections to motor field and armature windings. When triac conducts, normal line voltage less drop across triac and R5 is applied to motor. If firing of triac is delayed in each half of AC cycle, RMS voltage of motor is reduced and speed is correspondingly reduced. Feedback maintains torque at reduced speeds. Value of R5 in ohms is equal to 2 divided by rated RMS motor current in amperes and is 0.32 ohm for 6.5-A induction motor. Suitable for use with electric drills, where good torque is obtained down to about one-third of maximum speed.—"Circuit Applications for the Triac," Motorola, Phoenix, AZ, 1971, AN-466, p 7.

CLAMPED DARLINGTON CONTROL—Circuit uses MJE205 5-A transistor in Darlington configuration driving 2N5301 30-A transistor for passing 18 A to DC motor under logic control. CMOS gate connections are shown for energizing motor when logic is low and when logic is high.—A. Pshaenich, "Interface Techniques Between Industrial Logic and Power Devices," Motorola, Phoenix, AZ, 1975, AN-712A, p 19.

FREQUENCY CONTROLS SPEED—Circuit generates variable frequency between 10 and 300 Hz at constant voltage for changing speed of induction motor between theoretical limits of 300 and 9000 rpm without affecting maximum torque. Direct coupling between control and drive circuits is used; if motor noise affects control logic circuits, optoisolators should be used between control and drive sections. Article tells how circuit works and gives similar circuit using optical coupling.—T. Mazur, Unique Semiconductor Mix Controls Induction Motor Speed, *EDN Magazine,* Nov. 1, 1972, p 28–31.

2-hp THREE-PHASE INDUCTION—Speed is controlled by applying continuously variable DC voltage to VCO of control circuit for 750-VDC 7-A bridge inverter driving three sets of six Delco DTS-709 duolithic Darlingtons. Bridge inverter circuit for other two phases is identical to that shown for phase AA'. VCO output is converted to three-phase frequency varying from 5 Hz at 50 VDC to 60 Hz at 600 VDC for driving output Darlingtons. Optoisolators are used for base drive of three switching elements connected to high-voltage side of inverter.—"A 7A, 750 VDC Inverter for a 2 hp, 3 Phase, 480 VAC Induction Motor," Delco, Kokomo, IN, 1977, Application Note 60.

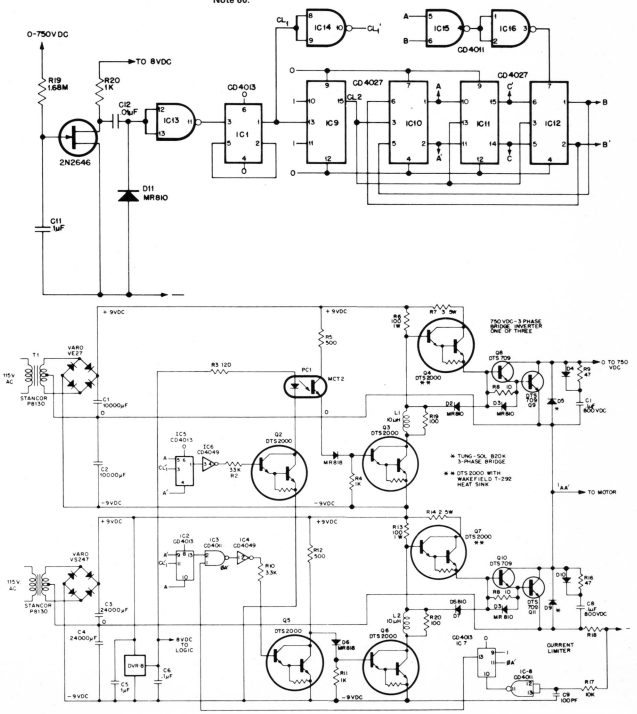

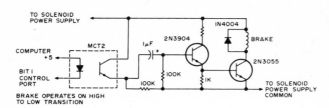

PULSED BRAKE—Transition from high (1) to low (0) at control port of microprocessor energizes brake solenoid of paper-tape reader in pulses lasting several microseconds, with time determined by size of capacitor used. Energizing of solenoid squeezes tape between top of solenoid and flat iron brake shoe that is attracted by solenoid. Unmarked resistor is 1K.—D. Hogg, The Paper Taper Caper, *Kilobaud*, March 1977, p 34–40.

SERIES-WOUND MOTOR—Provides both direction and speed control for fractional-horsepower series-wound or universal DC motors as long as motor current requirements are within SCR ratings. Q1-Q4, connected in bridge, are triggered in diagonal pairs. S1 determines which pair is turned on, to provide direction control. Pulse circuit is used to drive SCRs through T1 or T2. When C1 charges to breakdown voltage of zener D5, zener passes current to gate of SCR Q5 and turns it on. This discharges C1 through T1 or T2 to create desired triggering pulse. Q5 stays on for duration of half-cycle. R1 controls motor speed by changing time required to charge C1, thereby changing conduction angle of Q1-Q4 or Q2-Q3.—"Direction and Speed Control for Series, Universal and Shunt Motors," Motorola, Phoenix, AZ, 1976, AN-443.

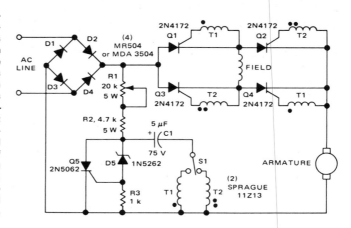

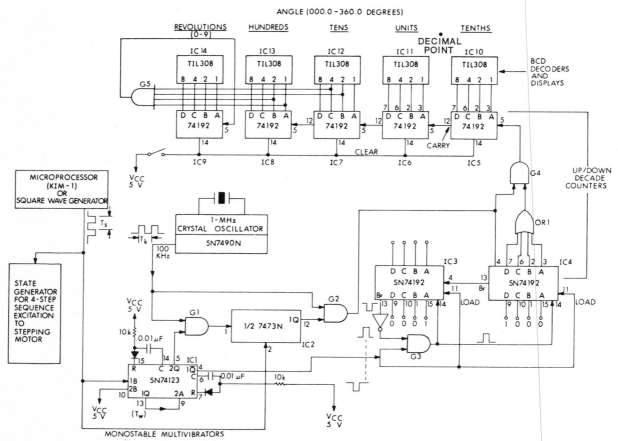

MOTOR STEP-ANGLE DISPLAY—Digital display circuit tracks stepper-motor shaft movements. Up/down decade counters read out four BCD digits as travel angle (000.0 to 360.0) in degrees and number of completed revolutions (0 to 9). Stepper under study is driven by state generator that produces high-current square-wave pulses under control of clock used for display, which can be external square-wave generator or clock output of microprocessor such as KIM-1. Power source for digital display is 5 V at 1.2 A. Applications include monitoring movements of incremental plotters, precision film camera drives, numerical control machines, and precision start-stop motions of fuel control rods in nuclear reactor.—H. Lo, Digital Display of Stepper Motor Rotation, *Computer Design*, April 1978, p 147–148 and 150–151.

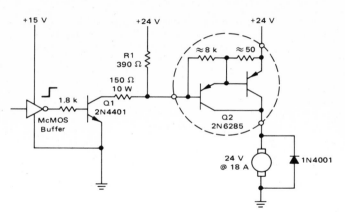

HIGH-LEVEL CMOS CONTROL—When output of CMOS buffer goes high, Q1 turns on and sinks 150-mA base current of power Darlington Q2, to activate motor load. Used in logic-controlled industrial applications.—A. Pshaenich, "Interface Techniques Between Industrial Logic and Power Devices," Motorola, Phoenix, AZ, 1975, AN-712A, p 18.

TELEFAX PHASING—Simple coincidence circuit provides reliable synchronization of Telefax machine in which 2500-Hz signal is generated by photoelectric scanning of paper placed on revolving drum. Circuit uses 7402 quad two-input NOR gate. If alternate connection enclosed in dashed line is not used, connect pin 8 to ground at pin 7. Q1 is S0014 silicon or equivalent. If relay contacts will handle motor voltage and current, they can be connected directly across points of test switch on machine, with switch left open for phasing circuit to work.—W. C. Smith, A Logic Circuit for Phasing the Telefax, QST, Nov. 1978, p 33–34.

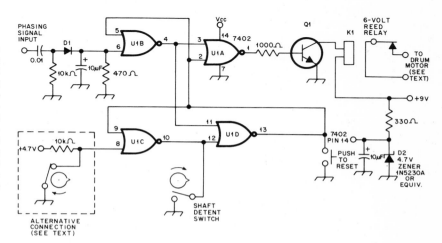

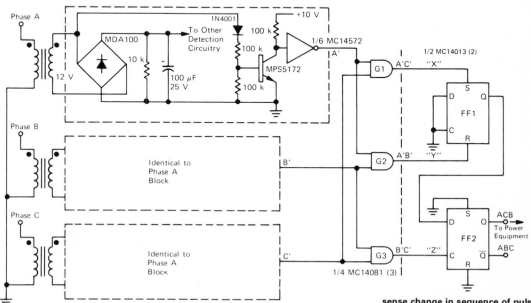

PHASE-REVERSAL DETECTOR—Used in three-phase applications in which direction of rotation of phases is critical, as in three-phase motors where reversal of two phases can provide disastrous reversal of motor. Line voltages are stepped down and isolated by control-type transformers. Each phase is half-wave rectified and shaped by 1N4001 diode and MPS5172 transistor, with additional shaping by MC14572 inverter. Shaped outputs of all three phases are combined in AND gates G1-G3 to give pulse outputs sequentially. D flip-flops are connected to sense change in sequence of pulses caused by reversal of one or more input phases. Flip-flop output can be used to trip relay or other protective device for removing air conditioner or other equipment from line before it is damaged.—T. Malarkey, "A Simple Line Phase-Reversal Detection Circuit," Motorola, Phoenix, AZ, 1975, EB-54.

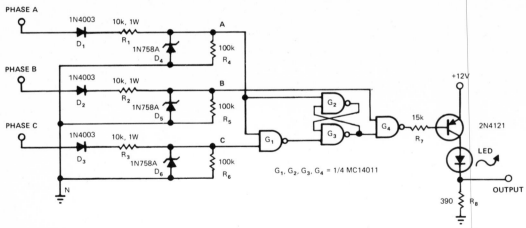

PHASE SEQUENCE DETECTOR—Circuit detects incorrect phase sequence of motor driving pump, compressor, conveyor, or other equipment that can be damaged by reverse rotation. Circuit also protects motor from phase loss that could cause rapid temperature rise and heat damage. LED is on when phasing is correct. For phase loss or incorrect sequence, output goes low and LED is dark. Diodes and zeners change sine waves for all phases to rectangular logic-level pulses that feed gates. When phases are correct, output of G_4 is train of rectangular pulses about 2.5 ns wide. Output is zero for incorrect sequences. Since leading edge of output pulse coincides with positive zero crossing of phase B, output pulses can be used to trigger SCR connected across phase B and driving relay-coil load. SCR then energizes relay only when sequence is correct.—H. Normet, Detector Protects 3-Phase-Powered Equipment, *EDN Magazine*, Aug. 5, 1978, p 78 and 80.

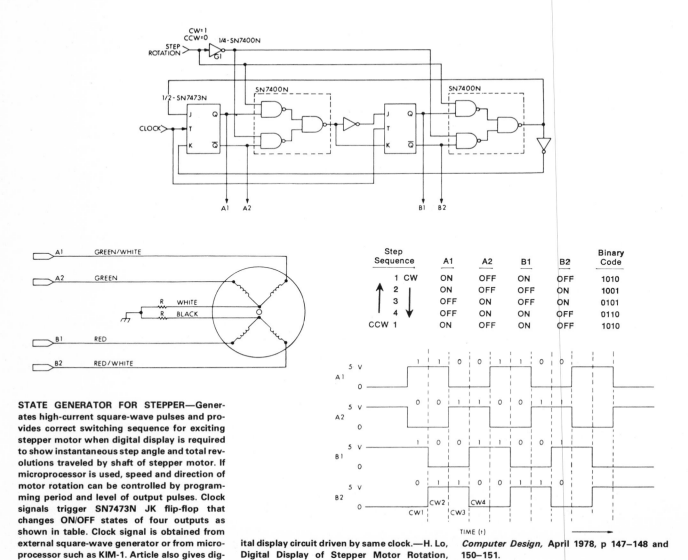

Step Sequence		A1	A2	B1	B2	Binary Code
1 CW		ON	OFF	ON	OFF	1010
2		ON	OFF	OFF	ON	1001
3		OFF	ON	OFF	ON	0101
4		OFF	ON	ON	OFF	0110
CCW 1		ON	OFF	ON	OFF	1010

STATE GENERATOR FOR STEPPER—Generates high-current square-wave pulses and provides correct switching sequence for exciting stepper motor when digital display is required to show instantaneous step angle and total revolutions traveled by shaft of stepper motor. If microprocessor is used, speed and direction of motor rotation can be controlled by programming period and level of output pulses. Clock signals trigger SN7473N JK flip-flop that changes ON/OFF states of four outputs as shown in table. Clock signal is obtained from external square-wave generator or from microprocessor such as KIM-1. Article also gives digital display circuit driven by same clock.—H. Lo, Digital Display of Stepper Motor Rotation, *Computer Design*, April 1978, p 147–148 and 150–151.

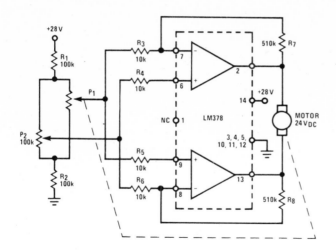

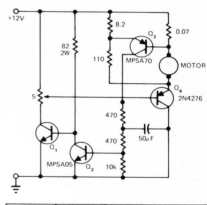

MOTOR VOLTAGE	TRANSISTOR VOLTAGE	Q_3 EMITTER VOLTS	LIMIT CURRENT
0	12	0	10A
6	6	0.42	16A
11.5	0.5	0.20	21.4A

24-VDC PROPORTIONAL SPEED CONTROL— National LM378 amplifier IC is basis for low-cost proportional speed controller capable of furnishing 700 mA continuously for such applications as antenna rotors and motor-controlled valves. Proportional control results from error signal developed across Wheatstone bridge R_1-R_2-P_1-P_2. P_1 is mechanically coupled to motor shaft as continuously variable feedback sensor. As motor turns, P_1 tracks movement and error signal becomes smaller and smaller; system stops when error voltage reaches 0 V.—"Audio Handbook," National Semiconductor, Santa Clara, CA, 1977, p 4-8–4-20.

STALLED-MOTOR PROTECTION—Modification of basic speed control circuit for small DC permanent-magnet motors provides maximum current limit under normal conditions and reduced current limit under stall conditions, to limit dissipation of series transistor Q_4 to safe value. When motor stalls, motor voltage falls, reducing voltage and motor current required to turn on Q_3 and thereby limiting stalled-motor current.—D. Zinder, Current Limit and Foldback for Small Motor Control, *EDN Magazine,* May 5, 1974, p 77 and 79.

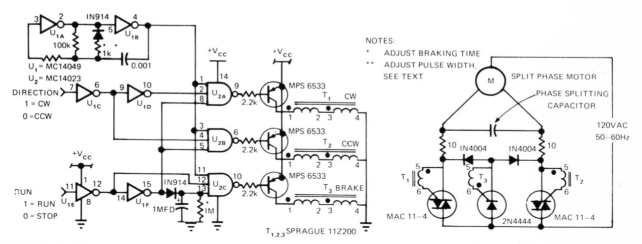

SPLIT-PHASE CONTROL WITH BRAKING—Use of CMOS logic to gate direction-controlling triacs and turn on SCR for braking provides low-cost switchless control of split-phase motor used in place of brush-type DC motor. Applications include control of ball valves and other throttling functions in process control. With shaft-position encoders, circuit generates feedback information. Overshoot and other stability problems are easily controlled by strong braking function. CMOS logic provides complete noise immunity. Oscillator pulse width is adjusted with 1K resistor in series with 1N914, and brake duration is controlled by 1-megohm resistor at input of U_{2C}. With values shown, brake is applied for about 1 s. Circuit works reliably on supply voltages of 5 to 15 V.—V. C. Gregory, Split-Phase Motor Control Accomplished with CMOS, *EDN Magazine,* Oct. 5, 1974, p 65–67.

CHAPTER 12

Music Circuits

Includes organ, piano, trombone, bell, theremin, bird-call, and other sound and music synthesizer circuits, along with circuits giving warble, fuzz, three-part harmony, reverberation, tremolo, attack, decay, rhythm, and other musical effects. Joystick control for music, active filters, contact-pickup preamp, metronomes, and tuning aids are also given.

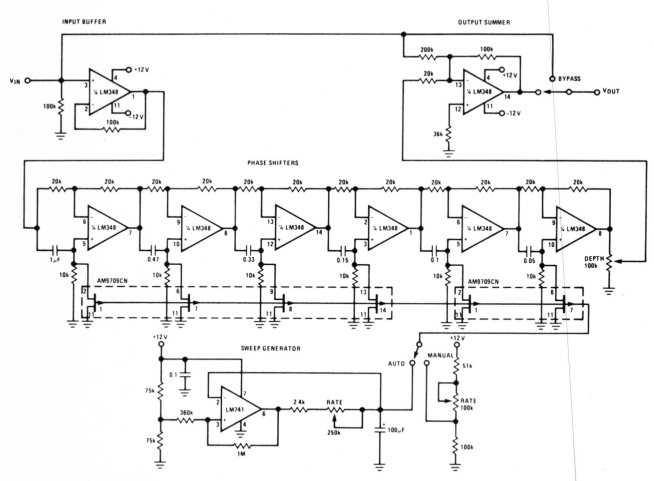

SIMULATION OF FLANGING—Sound-effect circuit sometimes called phase shifter simulates playing of two tape recorders having same material while varying speed of one by pressing on flange of tape reel. Resulting time delay causes some signals to be summed out of phase and canceled. Effect is that of rotating loudspeaker or of Doppler characteristic. Uses two LM348 quad opamps, two AM9709CN quad JFET devices, and one LM741 opamp. Phase-shift stages are spaced one octave apart from 160 to 3200 Hz in center of audio spectrum, with each stage providing 90° shift at its frequency. JFETs control phase shifters. Gate voltage of JFETs is adjusted from 5 V to 8 V either manually with foot-operated rheostat or automatically by LM741 triangle-wave generator whose rate is adjustable from 0.05 Hz to 5 Hz.—"Audio Handbook," National Semiconductor, Santa Clara, CA, 1977, p 5-10—5-11.

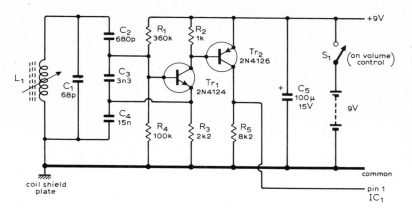

TUNING FOR EQUAL TEMPERAMENT—Instrument described enables anyone to tune such instruments as organ, piano, and harpsichord in equal temperament with accuracy approaching that of professional tuner. Only requirement is ability to hear beats between two tones sounded together. Master oscillator circuit shown generates 250.830 kHz for feeding to first of five ICs connected as programmable divider that provides 12 notes of an octave as 12 equal semitones differing from each other by factor of 1.0594. Article gives suitable power amplifier to fit along with divider connections and detailed instructions for construction, calibration, and use.—W. S. Pike, Digital Tuning Aid, *Wireless World*, July 1974, p 224–227.

TREMOLO CONTROL—National LM324 opamp connected as phase-shift oscillator operates at variable rate between 5 and 10 Hz set by speed pot. Portion of oscillator output is taken from depth pot and used to modulate ON resistance of two 1N914 diodes operating as voltage-controlled attenuators. Input should be kept below 0.6 V P-P to avoid undesirable clipping. Used for producing special musical effects.—"Audio Handbook," National Semiconductor, Santa Clara, CA, 1977, p 5-11–5-12.

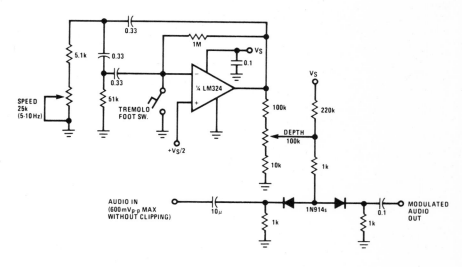

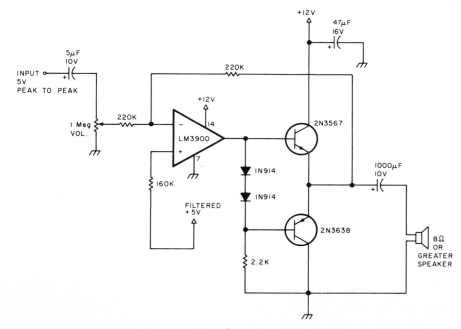

AUDIO FOR COMPUTER MUSIC—Wideband low-power audio amplifier was developed for use with DAC and low-pass active filter to create music with microprocessor.—H. Chamberlin, A Sampling of Techniques for Computer Performance of Music, *BYTE*, Sept. 1977, p 62–66, 68–70, 72, 74, 76–80, and 82–83.

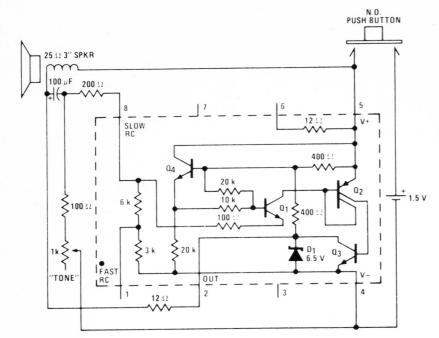

TROMBONE CIRCUIT—Unique arrangement for driving 25-ohm loudspeaker with National LM3909 IC operating from 1.5-V cell permits generation of slide tones resembling those of trombone. Operation is based on use of voltage generated by resonant motion of loudspeaker voice coil as major positive feedback for IC. Loudspeaker is mounted in roughly cubical box having volume of about 64 in³, with one end of box arranged to slide in and out like piston. Positioning of piston and operation of pushbutton permit playing reasonable semblance of simple tune. IC, loudspeaker, and battery are mounted on piston, with 2½-in length of ⁵⁄₁₆-in tubing provided to bleed air in and out as piston is moved, without affecting resonant frequency. Frequency of oscillator becomes equal to resonant frequency of enclosure.—"Linear Applications, Vol. 2," National Semiconductor, Santa Clara, CA, 1976, AN-154, p 6.

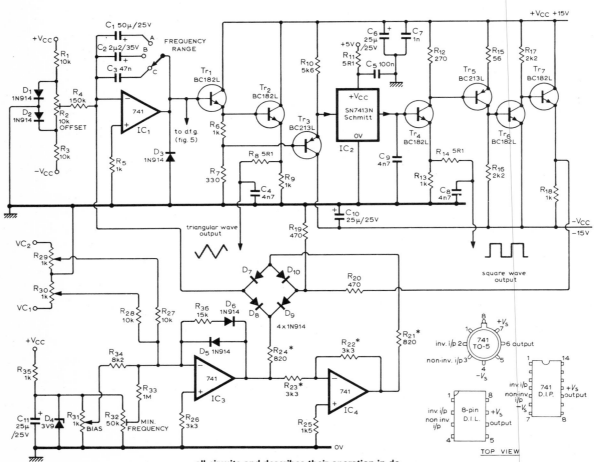

VCO SOUND SYNTHESIZER—Developed for use in instrument capable of duplicating variety of sounds ranging from bird distress calls and engine noises to spoken words and wide variety of musical instruments. Three-part article gives all circuits and describes their operation in detail. Heart of oscillator is triangle and square-wave generator built around IC Schmitt trigger. Ramp rate and operating frequency are varied by changing drive voltage or gain of integrator. Similar VCO in synthesizer also produces sine, pulse, and ramp waveforms.—T. Orr and D. W. Thomas, Electronic Sound Synthesizer, *Wireless World,* Part 1—Aug. 1973, p 366–372 (Part 2—Sept. 1973, p 429–434; Part 3—Oct. 1973, p 485–490).

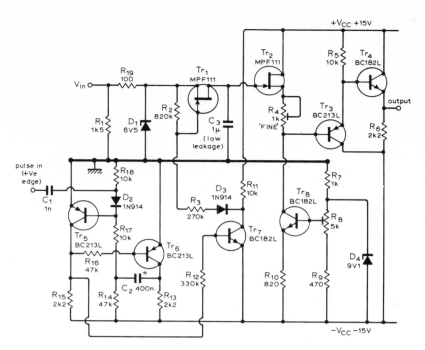

ANALOG MEMORY—Used in synthesizer for generating wide variety of musical and other sounds, to provide constant control signal for sounds requiring long fadeout. Positive input pulse initiates sampling of analog signal for preset time, with signal being held for unspecified period. Input voltage range is from about −0.5 V to +6.5 V, being deliberately limited by D_1. Three-part article describes operation in detail and gives all other circuits used in synthesizer.—T. Orr and D. W. Thomas, Electronic Sound Synthesizer, *Wireless World,* Part 3—Oct. 1973, p 485–490 (Part 1—Aug. 1973, p 366–372; Part 2—Sept. 1973, p 429–434).

TREMOLO AMPLIFIER—Provides amplitude modulation at subaudio rate (usually between 5 and 15 Hz) of audio-frequency input signal. Uses National LM389 array having three transistors along with power amplifier. Transistors form differential pair having active current-source tail to give output proportional to product of two input signals. Gain control pot is adjusted for desired tremolo depth. Interstage RC network forms 160-Hz high-pass filter, requiring that tremolo frequency be less than 160 Hz.— "Audio Handbook," National Semiconductor, Santa Clara, CA, 1977, p 4-33–4-37.

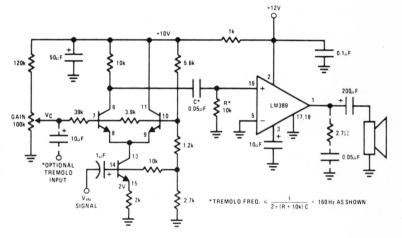

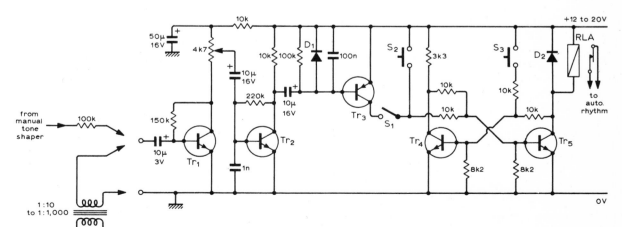

AUTOMATIC REMOTE RHYTHM CONTROL— When added to electronic organ, circuit is activated by audio signal from lower manual or pedal, to initiate start of rhythm accompaniment. High-impedance input connection through 100K is made to toneshaper output, and transformer connection is used with electromechanical Hammond organ. Transistor and diode types are not critical. If S_1 is closed, current passes through to Tr_5 and triggers bistable that pulls in relay. S_2 and S_3 are used for manual start and stop of rhythm.—K. B. Sorensen, Touch Start of Automatic Rhythm Device, *Wireless World,* Oct. 1974, p 381.

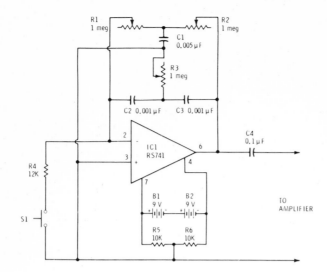

MUSICAL BELLS—Opamp connected as active filter simulates attack followed by gradual decay as produced when bell or tuning fork is struck. Filter portion of circuit uses twin-T network adjusted so active filter breaks into oscillation when slight external disturbance is introduced by closing S1 momentarily. Circuit feeds external audio amplifier and loudspeaker for converting ringing frequency into audible sound. Set R3 just below oscillation point. R1 and R2 can be adjusted to give sounds of other musical instruments, such as drums, bamboo, and triangles.—F. M. Mims, "Electronic Music Projects, Vol. 1," Radio Shack, Fort Worth, TX, 1977, 2nd Ed., p 71–80.

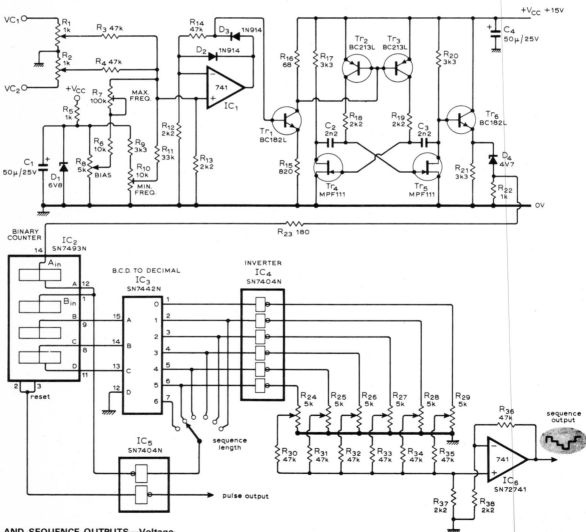

PULSE AND SEQUENCE OUTPUTS—Voltage-controlled oscillator produces sequence of steps, with amplitude of each step individually controllable up to maximum of six steps. Circuit also generates series of pulses having 1:1 mark-space ratio, each coincident with leading edge of a step. Pair of summing inputs controls oscillator, with exponential frequency-voltage relationship extending in one range from subsonic frequencies to over 20 kHz. Used in sound synthesizer described in three-part article that gives all circuits and operating details. Applications include synthesizing sounds ranging from bird distress calls and engine noises to spoken words and wide variety of musical instruments.—T. Orr and D. W. Thomas, Electronic Sound Synthesizer, *Wireless World*, Part 2—Sept. 1973, p 429–434 (Part 1—Aug. 1973, p 366–372; Part 3—Oct. 1973, p 485–490).

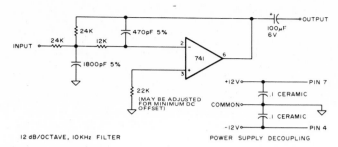

12 dB/OCTAVE, 10KHz FILTER

POWER SUPPLY DECOUPLING

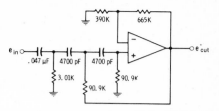

10-kHz LOW-PASS FILTER—Suitable for use at both input and output of A/D-D/A converter in digital audio system for synthesizing speech or music. Serves for smoothing steps of output waveform and suppressing background noise on output when small signals are being processed with 8-bit linear encoding.—T. Scott, Digital Audio, *Kilobaud,* May 1977, p 82–86.

327-Hz HIGH-PASS—Developed to make third harmonic of 130.81 Hz (C3 note) minimum of 30 dB stronger than fundamental, to give sawtooth output for use in electronic music system. Design uses third-order filter with 3-dB dips in response. Opamp can be 741.—D. Lancaster, "Active-Filter Cookbook," Howard W. Sams, Indianapolis, IN, 1975, p 192.

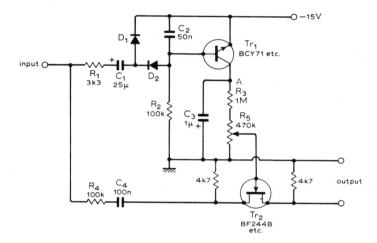

PIANO MUSIC FROM ORGAN—Simple add-on circuit for electronic organ attenuates output of oscillator exponentially to zero in manner suitable for mimicking waveform of piano. Circuit is self-triggering, so exponential decay starts only when output of multivibrator is applied; this eliminates need for extra contacts on keyboard. With no input, Tr_1 is on and point A is at supply voltage. Input signal turns Tr_1 off, discharging C_3 through R_3 and R_5. Voltage across C_3 controls gate of FET, with R_5 being adjusted so FET just switches off when C_3 is fully discharged. Tr_1 then conducts and C_3 charges rapidly, to permit fast piano playing.—C. J. Outlaw, Electronic Organ to Piano, *Wireless World,* Feb. 1975, p 94.

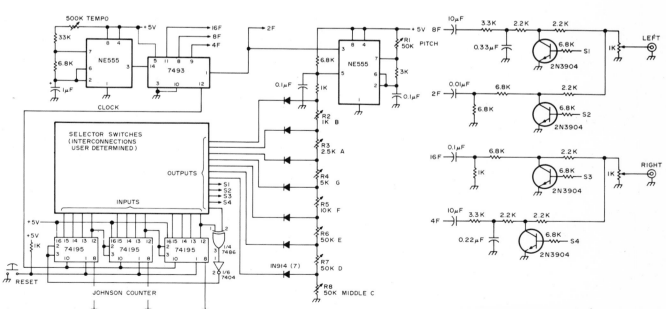

RANDOM MUSIC—Uses Johnson counter as special shift register producing almost random bit patterns of 18 to 3255 12-bit words under control of clock operating in range of about 1–10 Hz. Oscillator (upper right) uses NE555 as voltage-controlled square-wave generator playing one of eight musical notes (C, D, E, F, G, A, B, or C), depending on state of seven note-selector lines coming from selector switches. Oscillator is divided down in frequency by three-stage ripple counter to provide four octaves of range. R1-R8 serve for tuning each note to pitch. Outputs 8F and 2F are paired to drive left input of stereo amplifier, while outputs 16F and 4F are similarly paired for right channel. Article covers construction, tune-up, and creation of pleasing musical sequences.—D. A. Wallace, The Sound of Random Numbers, 73 *Magazine,* Feb. 1976, p 60–64.

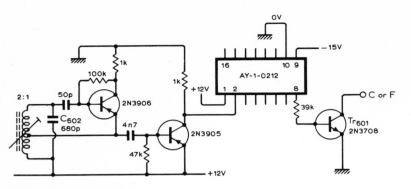

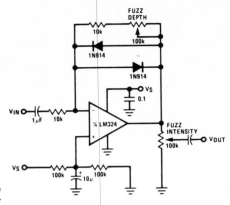

PIANO TONE GENERATOR—RF oscillator combined with General Instrument AY-1-0212 IC master tone generator replaces 12 conventional RC oscillators otherwise required in electronic piano. Frequencies generated are within 0.1% of equal-temperament scale, so piano will work well without being tuned. Three-part article gives all circuits and construction details for simple portable touch-sensitive electronic piano.—G. Cowie, Electronic Piano Design, *Wireless World*, Part 3, May 1974, p 143–145.

FUZZ CIRCUIT—Two diodes in feedback path of LM324 opamp create musical-instrument effect known as fuzz by limiting output voltage swing to ±0.7 V. Resultant square wave contains chiefly odd harmonics, resembling sounds of clarinet. Fuzz depth pot controls level at which clipping begins, and fuzz intensity pot controls output level.—"Audio Handbook," National Semiconductor, Santa Clara, CA, 1977, p 5-11.

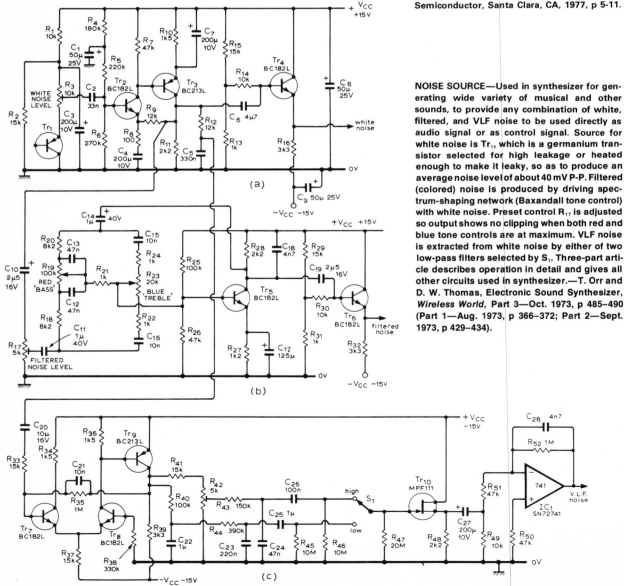

NOISE SOURCE—Used in synthesizer for generating wide variety of musical and other sounds, to provide any combination of white, filtered, and VLF noise to be used directly as audio signal or as control signal. Source for white noise is Tr_1, which is a germanium transistor selected for high leakage or heated enough to make it leaky, so as to produce an average noise level of about 40 mV P-P. Filtered (colored) noise is produced by driving spectrum-shaping network (Baxandall tone control) with white noise. Preset control R_{17} is adjusted so output shows no clipping when both red and blue tone controls are at maximum. VLF noise is extracted from white noise by either of two low-pass filters selected by S_1. Three-part article describes operation in detail and gives all other circuits used in synthesizer.—T. Orr and D. W. Thomas, Electronic Sound Synthesizer, *Wireless World*, Part 3—Oct. 1973, p 485–490 (Part 1—Aug. 1973, p 366–372; Part 2—Sept. 1973, p 429–434).

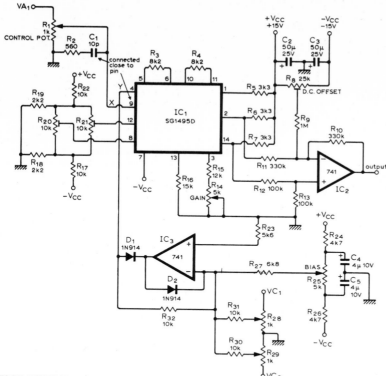

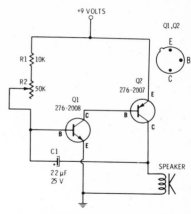

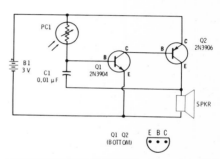

VOLTAGE-CONTROLLED AMPLIFIER—Gain is linearly controlled by sum of input control voltages and a bias voltage, to provide amplitude modulation as required for synthesizer used to generate wide variety of sounds. Heart of circuit is linear four-quadrant multiplier IC. Output is taken between two load resistors, with differential amplifier IC_2 removing common-mode signal. Article describes operation in detail and gives all other circuits of synthesizer, along with procedure for aligning preset controls R_8, R_{14}, R_{20}, and R_{21}.—T. Orr and D. W. Thomas, Electronic Sound Synthesizer, *Wireless World*, Part 2—Sept. 1973, p 429–434 (Part 1—Aug. 1973, p 366–372; Part 3—Oct. 1973, p 485–490).

CLICKING METRONOME—Basic lamp-flashing circuit is used to produce sharp click in loudspeaker each time Q2 is turned on by RC oscillator Q1. R2 adjusts repetition rate over range of 20–280 beats per minute. Changing value of C1 varies tone of clicks.—F. M. Mims, "Transistor Projects, Vol. 1," Radio Shack, Fort Worth, TX, 1977, 2nd Ed., p 33–39.

LIGHT-SENSITIVE THEREMIN—Tone of loudspeaker increases and decreases in frequency as flashlight is moved in vicinity of photocell in darkened room. Use Radio Shack 276-116 cadmium sulfide photocell. Cell resistance decreases with light, increasing frequency of audio oscillator. Continuously changing frequency resembles that produced by hand-controlled theremin.—F. M. Mims, "Electronic Music Projects, Vol. 1," Radio Shack, Fort Worth, TX, 1977, 2nd Ed., p 91–95.

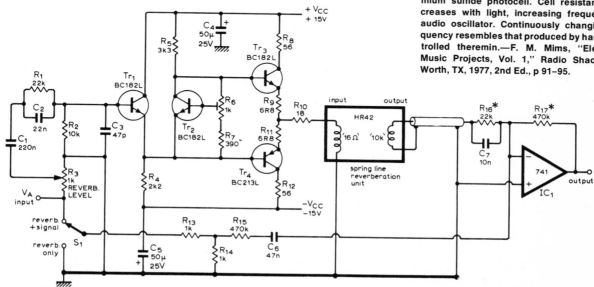

REVERBERATION—Used in sound synthesizer developed for generating wide variety of musical and other sounds. Four-transistor driver feeds spring-type reverberation unit at up to about 4 kHz, with switch giving choice of reverberation only or reverberation combined with input signal at V_A. Amount of reverberation can be controlled manually with R_3 or automatically with voltage-controlled amplifier or voltage-controlled filter of synthesizer. Three-part article gives all circuits and describes operation in detail.—T. Orr and D. W. Thomas, Electronic Sound Synthesizer, *Wireless World*, Part 2—Sept. 1973, p 429–434 (Part 1—Aug. 1973, p 366–372; Part 3—Oct. 1973, p 485–490).

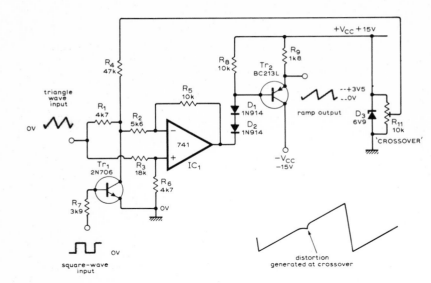

RAMP FUNCTION—Circuit combines triangle and square-wave inputs from VCO in differential amplifier having switched gain, to generate ramp function for use with variety of other waveforms in sound synthesizer designed for duplicating wide variety of sounds. Three-part article gives all circuits and operating details.—T. Orr and D. W. Thomas, Electronic Sound Synthesizer, *Wireless World,* Part 1—Aug. 1973, p 366–372 (Part 2—Sept. 1973, p 429–434; Part 3—Oct. 1973, p 485–490).

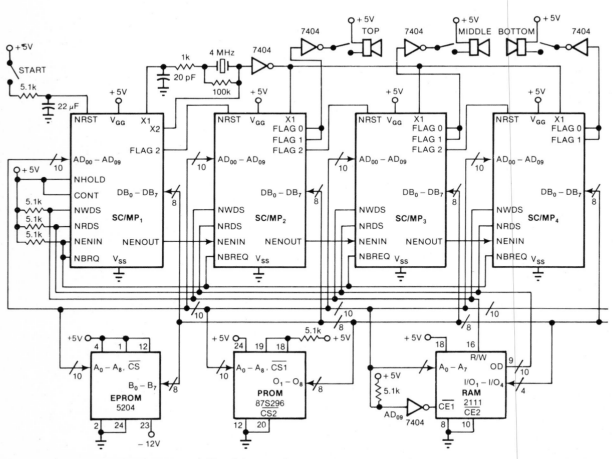

THREE-PART HARMONY—Four SC/MP microprocessors, one serving as conductor and three as instrumentalists, generate multiple parts for harmony feeding common loudspeaker system. Microprocessors have paralleled address and data buses, with 4K RAM connecting to lowest 4 bits of data bus. Each microprocessor is supplied with list of notes by note number and note lengths as part of software. At end of each basic note length, SC/MP₁ checks each other processor to see if it is time to proceed to next note in list. If it is, next note is played by other processors until signaled by conductor via memory. Article gives software listing.—T. Doone, Quartet of SC/MP's Plays Music for Trios, *EDN Magazine,* Sept. 20, 1978, p 57–58 and 60.

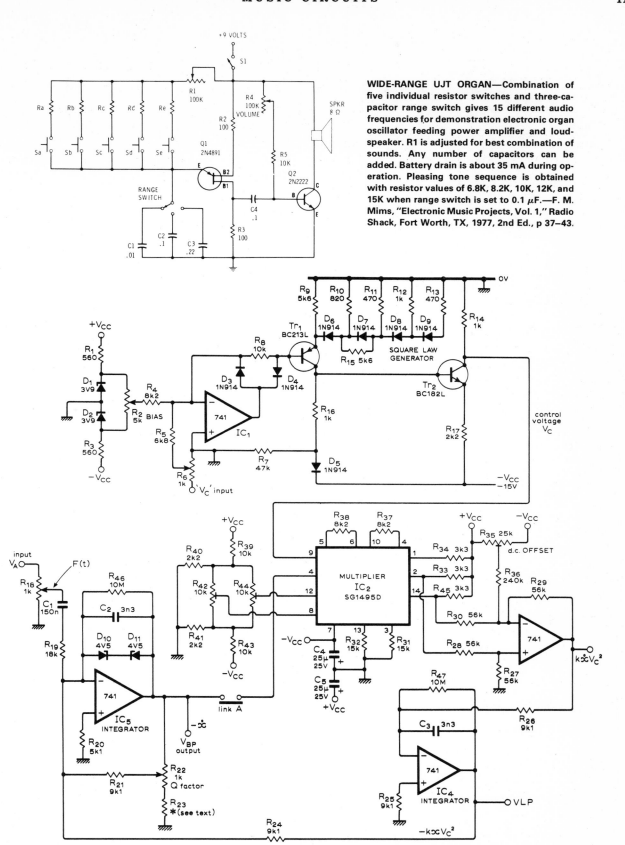

WIDE-RANGE UJT ORGAN—Combination of five individual resistor switches and three-capacitor range switch gives 15 different audio frequencies for demonstration electronic organ oscillator feeding power amplifier and loudspeaker. R1 is adjusted for best combination of sounds. Any number of capacitors can be added. Battery drain is about 35 mA during operation. Pleasing tone sequence is obtained with resistor values of 6.8K, 8.2K, 10K, 12K, and 15K when range switch is set to 0.1 μF.—F. M. Mims, "Electronic Music Projects, Vol. 1," Radio Shack, Fort Worth, TX, 1977, 2nd Ed., p 37–43.

VOLTAGE-CONTROLLED FILTER—Used in elaborate sound synthesizer developed for generating wide variety of sounds. Serves as bandpass filter for which resonant frequency is linearly proportional to sum of input control voltages and a bias voltage. Can also be used as notch filter or as spectrum analyzer. Three-part article describes operation in detail and gives all other circuits of synthesizer. Supply voltages are 15 V, with polarity as indicated.—T. Orr and D. W. Thomas, Electronic Sound Synthesizer, *Wireless World,* Part 2—Sept. 1973, p 429–434 (Part 1—Aug. 1973, p 366–372; Part 3—Oct. 1973, p 485–490).

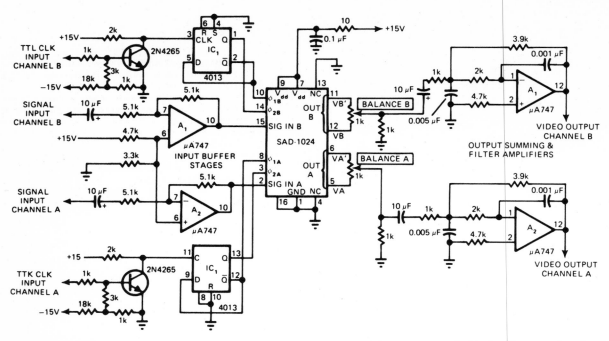

CCD DELAY FOR SPECIAL EFFECTS—Basic bucket-brigade device incorporated in Reticon Corp. SAD-1024 charge-coupled-device delay line can synthesize such interesting audio-system delay effects as reverberation enhancement, chorus, and vibrato generation. Other applications include speech compression and voice scrambling. Evaluation circuit shown was developed by manufacturer. Input clock frequency is 200 kHz, and signal input is 5-kHz sine wave. Article describes operation of evaluation circuit in detail and presents variety of practical applications.—R. R. Buss, CCD's Improve Audio System Performance and Generate Effects, *EDN Magazine*, Jan. 5, 1977, p 55–61.

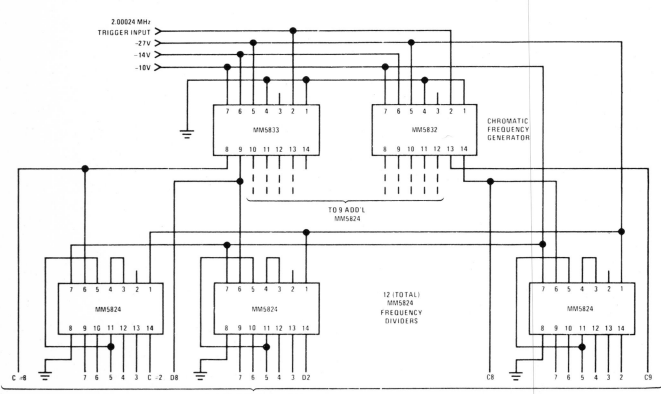

ORGAN TONE GENERATOR—National MM5832 and MM5833 chromatic frequency generators are used with 12 MM5824 frequency dividers to generate 85 musical frequencies fully spanning equal-tempered octave. Can also be used as celeste or chorus tone generator and as electronic music synthesizer. Square-wave input for organ is 2.00024 MHz but can be as low as 7 kHz for other applications.—"MOS/LSI Databook," National Semiconductor, Santa Clara, CA, 1977, p 3-11–3-13.

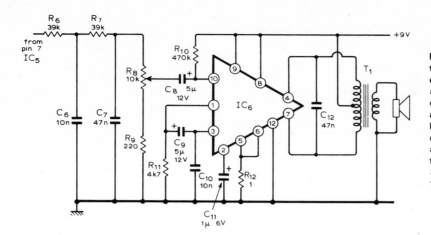

PIANO-TUNING AMPLIFIER—Used with battery-powered digital tuning aid that provides 12 equal semitones of octave, between 261.6625 and 493.8833 Hz, for equal-temperament tuning of such keyboard instruments as organ, piano, and harpsichord. IC used is part of RCA amplifier kit KC-4003 which includes T_1 and other discrete components. Article gives circuits for oscillator and programmable divider, along with instructions for construction, calibration, and use.—W. S. Pike, Digital Tuning Aid, *Wireless World*, July 1974, p 224–227.

WARBLER—One 555 timer is connected as low-frequency square-wave generator that modulates second timer producing higher-frequency tone, to give warbling tone that can be varied with R1 and R4 to simulate siren or songs of certain birds. Will operate over supply range of 4.5–18 V. Use 8-ohm miniature loudspeaker, with optional volume control R7 in series.—F. M. Mims, "Electronic Music Projects, Vol. 1," Radio Shack, Fort Worth, TX, 1977, 2nd Ed., p 29–35.

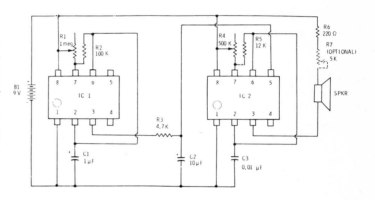

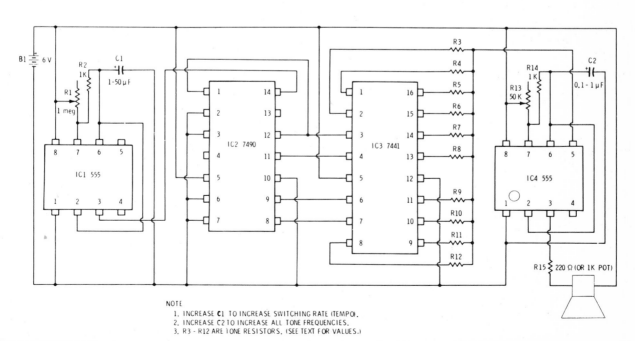

NOTE
1. INCREASE C1 TO INCREASE SWITCHING RATE (TEMPO).
2. INCREASE C2 TO INCREASE ALL TONE FREQUENCIES.
3. R3 - R12 ARE TONE RESISTORS. (SEE TEXT FOR VALUES.)

MUSIC SYNTHESIZER—555 timer is used as clock to set beat, adjustable with R1. Beat pulses drive flip-flop chain in 7490, which provides running total in binary format to 7441 1-of-10 decoder for conversion to decimal output. Each of ten outputs feeds through tone-controlling resistor to modulation input terminal of another 555 used as voltage-controlled AF oscillator feeding loudspeaker. Values used for resistors determine frequencies of ten notes that are generated in sequence repeatedly. R3-R12 can be 1000-ohm pots, so ten-note tune being played can be easily changed.—F. M. Mims, "Electronic Music Projects, Vol. 1," Radio Shack, Fort Worth, TX, 1977, 2nd Ed., p 61–70.

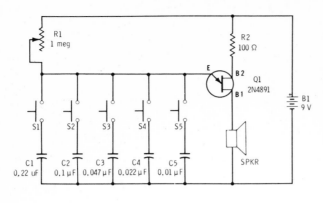

BASIC UJT ORGAN—Switches give choice of five audio frequencies for simple transistor oscillator driving loudspeaker. Adjust R1 for best combination of sounds. Capacitor values can be changed to give other frequencies. Ideal for classroom demonstrations.—F. M. Mims, "Electronic Music Projects, Vol. 1," Radio Shack, Fort Worth, TX, 1977, 2nd Ed., p 37–43.

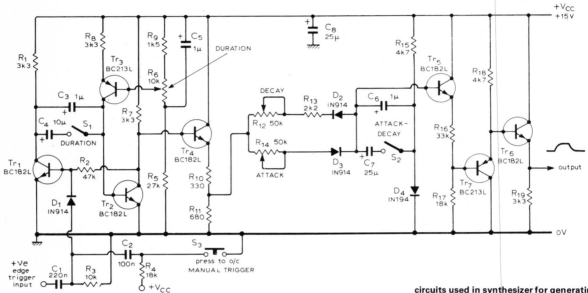

ATTACK/DECAY—Waveform generator produces approximate rectangular waveform having exponential rise (attack) and exponential decay, initiated either by manual trigger or electronic signal derived from other circuits of sound synthesizer. All characteristics of waveform are arbitrarily variable. Three-part article describes circuit operation and gives all other circuits used in synthesizer for generating wide variety of musical and other sounds.—T. Orr and D. W. Thomas, Electronic Sound Synthesizer, *Wireless World,* Part 3—Oct. 1973, p 485–490 (Part 1—Aug. 1973, p 366–372; Part 2—Sept. 1973, p 429–434).

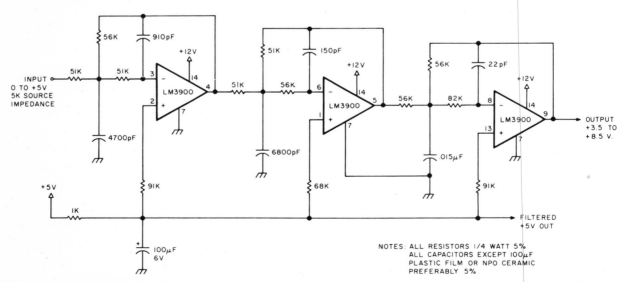

LOW-PASS WITH 3-kHz CUTOFF—Used in computer music system to suppress high-fidelity distortion resulting from steps in sine-wave output of DAC. Article covers complete computer synthesis of music by microprocessor and gives frequency table, program for generating four simultaneous musical voices, and song table for encoding "The Star Spangled Banner" in four-part harmony, using 5 bytes per musical event.—H. Chamberlin, A Sampling of Techniques for Computer Performance of Music, *BYTE,* Sept. 1977, p 62–66, 68–70, 72, 74, 76–80, and 82–83.

ATTACK-DECAY GENERATOR—Designed for polytonic electronic music system handling more than one note at a time. Each note to be controlled is sent through voltage-controlled amplifier (VCA) whose gain is set by charge on capacitor. Attack is changed by varying charging rate. Discharge rate sets decay of individual note. To avoid having separate adjustment pot for each VCA, duty-cycle modulation is used to change charging current through resistors. Attack pulses are generated by upper three inverters forming variable-symmetry astable MVBR. Decay pulses are generated by lower three inverters connected as half-mono MVBR. Additional half-monos can be added as needed

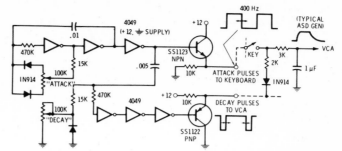

for percussion, snubbing, and other two-step decay effects.—D. Lancaster, "CMOS Cookbook," Howard W. Sams, Indianapolis, IN, 1977, p 231–232.

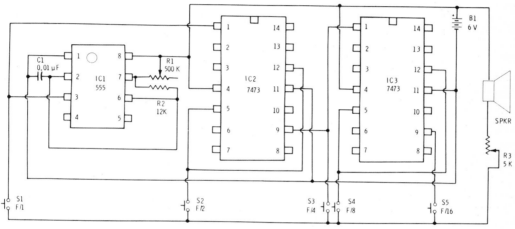

FOUR-OCTAVE ORGAN—Two 7473 dual flip-flops provide four frequency dividers for 555 timer connected as master tone generator. S1 gives fundamental frequency, and each succeeding switch gives tones precisely one octave lower. Four organ applications, pushbutton switches are added to timer circuit for switching frequency-controlling capacitors or resistors to give desired variety of notes.—F. M. Mims, "Electronic Music Projects, Vol. 1," Radio Shack, Fort Worth, TX, 1977, 2nd Ed., p 45–53.

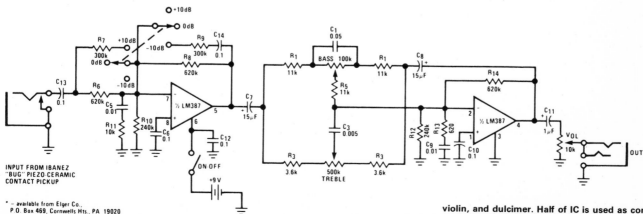

PREAMP FOR ACOUSTIC PICKUP—National LM387 dual opamp provides switchable gain choice of ±10 dB along with bass/treble tone control and volume control. Used with flat-response piezoceramic contact pickup for acoustic stringed musical instruments such as guitar, violin, and dulcimer. Half of IC is used as controllable gain stage, and other half is used as active two-band tone-control block.—"Audio Handbook," National Semiconductor, Santa Clara, CA, 1977, p 5-12.

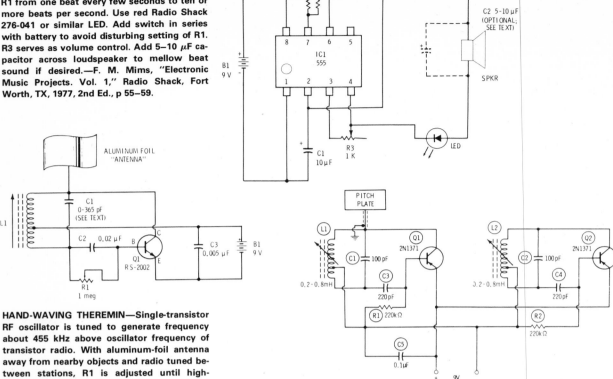

AUDIBLE/VISIBLE METRONOME—Produces uniformly spaced beats in synchronism with flashes of LED, at rate that can be adjusted with R1 from one beat every few seconds to ten or more beats per second. Use red Radio Shack 276-041 or similar LED. Add switch in series with battery to avoid disturbing setting of R1. R3 serves as volume control. Add 5–10 μF capacitor across loudspeaker to mellow beat sound if desired.—F. M. Mims, "Electronic Music Projects. Vol. 1," Radio Shack, Fort Worth, TX, 1977, 2nd Ed., p 55–59.

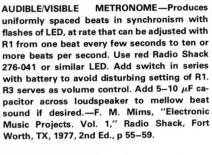

HAND-WAVING THEREMIN—Single-transistor RF oscillator is tuned to generate frequency about 455 kHz above oscillator frequency of transistor radio. With aluminum-foil antenna away from nearby objects and radio tuned between stations, R1 is adjusted until high-pitched tone is heard from radio. Now, as hand is brought toward and away from foil antenna, wailing sounds are produced. With practice, musician can produce recognizable melodies by vibrating hand. Primary controls of frequency are C1 (10–365 pF broadcast radio tuning capacitor) and adjustable antenna coil L1 (Radio Shack 270-1430). Radio can be up to 15 feet away from theremin. Rotate radio for maximum pickup from L1.—F. M. Mims, "Electronic Music Projects, Vol. 1," Radio Shack, Fort Worth, TX, 1977, 2nd Ed., p 81–89.

THEREMIN—Two transistor oscillator stages generate separate low-power RF signal in broadcast band, for pickup by AM broadcast receiver. Movement of hand toward or away from metal pitch plate varies frequency of Q1, making audio output of receiver vary correspondingly as beat frequency changes. Both circuits are Hartley oscillators, using Miller 9012 or equivalent slug-tuned coils. To adjust initially, place next to radio and set tuning slug of L1 about two-thirds out of its winding. Set slug of L2 about one-third out of its winding. Tune radio until either oscillator signal is heard. Signal can be identified by whistle if on top of broadcast station or by quieting of background noise if between stations. Adjust slugs so whistle is heard at desired location of quieting signal. Pitch of whistle should change now as hand is brought near pitch plate.—J. P. Shields, "How to Build Proximity Detectors & Metal Locators," Howard W. Sams, Indianapolis, IN, 2nd Ed., 1972, p 154–156.

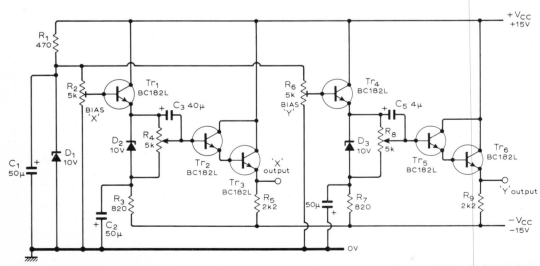

JOYSTICK CONTROL—Mechanically controlled voltage source generates two independent control voltages, proportional to stick position, to serve as one of controls for elaborate sound synthesizer used for generating wide variety of musical and other sounds. Three-part article describes circuit operation and gives all other circuits used in synthesizer.—T. Orr and D. W. Thomas, Electronic Sound Synthesizer, *Wireless World,* Part 3—Oct. 1973, p 485–490 (Part 1—Aug. 1973, p 366–372; Part 2—Sept. 1973, p 429–434).

CHAPTER 13
Phonograph Circuits

Includes RIAA-equalized preamps for all types of mono and stereo phono pickups, along with power amplifiers, tone controls, rumble and scratch filters, and test circuits.

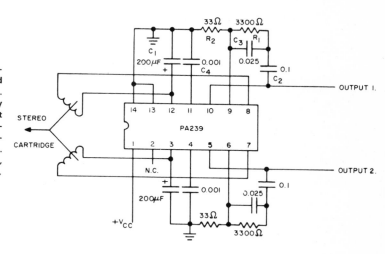

MAGNETIC-CARTRIDGE PREAMP—Uses Signetics PA239 dual low-noise amplifier designed specifically for low-level low-noise applications. Stereo channel separation at 1 kHz is typically 90 dB, and total harmonic distortion without feedback is 0.5%. Circuit matches amplifier response with RIAA recording characteristic. Supply voltage can be between 9 and 15 V at 22 mA. Article gives design equations.—A. G. Ogilvie, Construct a Magnetic-Cartridge Preamp, *Audio,* June 1974, p 40 and 42.

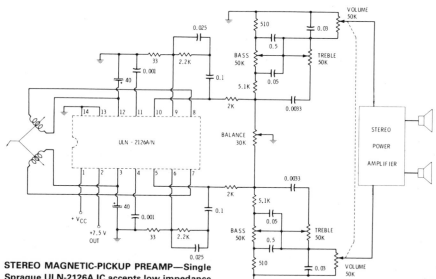

STEREO MAGNETIC-PICKUP PREAMP—Single Sprague ULN-2126A IC accepts low impedance of magnetic cartridge and provides up to 2-W output power for driving commercial stereo power amplifier. Circuit includes balance control and all tone controls along with ganged volume control. Values shown give proper equali-zation for playback of records.—E. M. Noll, "Linear IC Principles, Experiments, and Projects," Howard W. Sams, Indianapolis, IN, 1974, p 237 and 242.

NEW RIAA NETWORK—Values of R7 and C2 have been changed as shown in standard network for phonograph playback equalization. Tantalum electrolytic rated at least 20 V is recommended for C2. Network can also be used as inverse RIAA equalizer for testing preamps, with signal applied to terminal 2 and output to preamp taken from terminal 1. New standard extends playback equalization to 20,000 Hz and specifies that equalization be 3 dB down from previous standard at 20 Hz, with rolloff at 6 dB per octave below 20 Hz.—W. M. Leach, New RIAA Feedback Network, *Audio,* March 1978, p 103.

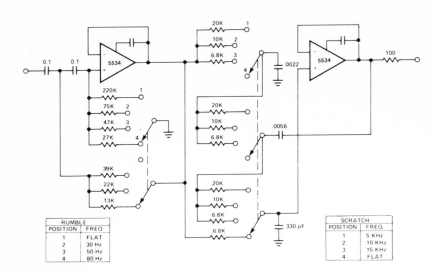

RUMBLE	
POSITION	FREQ.
1	FLAT
2	30 Hz
3	50 Hz
4	80 Hz

SCRATCH	
POSITION	FREQ.
1	5 KHz
2	10 KHz
3	15 KHz
4	FLAT

RUMBLE/SCRATCH FILTER—Used after pre-amp in high-quality audio system to improve reproduction of phonograph records. Two-pole Butterworth design has switchable breakpoints providing any desired degree of filtering.— "Signetics Analog Data Manual," Signetics, Sunnyvale, CA, 1977, p 638–639.

SCRATCH FILTER—Provides passband gain of 1 and corner frequency of 10 kHz for rolling off excess high-frequency noise appearing as hiss, ticks, and pops from worn records. Design procedure is given.—"Audio Handbook," National Semiconductor, Santa Clara, CA, 1977, p 2-49–2-52.

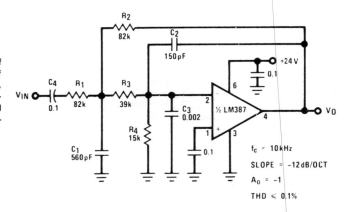

$f_c = 10 \text{ kHz}$

$\text{SLOPE} = -12 \text{ dB/OCT}$

$A_0 = -1$

$\text{THD} \leq 0.1\%$

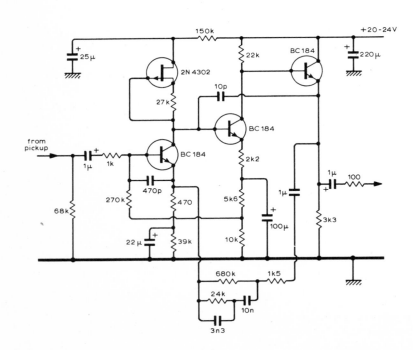

RIAA PREAMP—Low-noise circuit (below −70 dB referred to 5-mV input from pickup) has high overload capability and low distortion (below 0.05% intermodulation at 2 VRMS output). Arrangement of first stage gives improved transient reponse over usual feedback pair. Second stage provides gain of 10.—S. F. Bywaters, RIAA-Equalized Pre-Amplifier, *Wireless World*, Dec. 1974, p 503.

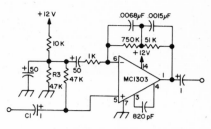

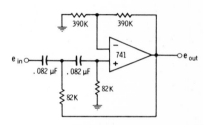

INPUT BUFFER FOR PREAMP—Used between cartridge and preamp of each stereo channel to make comparison testing of phonograph preamps more nearly independent of cartridge and cable capacitances. Buffer terminates cartridge in 47K in parallel with C1. Buffer can then serve as sonic reference for comparison with preamps for which input impedance is unknown. Article tells how to determine correct value of C1 for cartridge used and covers preamp test procedures in detail.—T. Holman, New Tests for Preamplifiers, *Audio*, Feb. 1977, p 58, 60, 62, and 64.

MAGNETIC-CARTRIDGE PREAMP—Uses dual opamp for stereo, other half of which is connected exactly the same but with connections to pin numbers changed to those in parentheses: 6 (5), 5 (8), 3 (11), 4 (10), and 1 (13).—Circuits, *73 Magazine*, Sept. 1973, p 143.

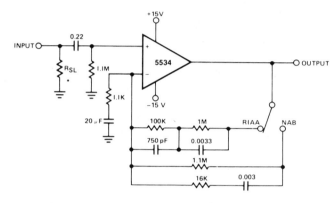

EQUALIZED PREAMP—Low-frequency boost is provided by inductance of magnetic cartridge, acting with RC network to approximate theoretical RIAA or NAB compensation as determined by position of compensation switch. Input resistor is selected to provide specified loading for cartridge. Output noise is about 0.8 mVRMS with input shorted.—"Signetics Analog Data Manual," Signetics, Sunnyvale, CA, 1977, p 638–639.

20-Hz HIGH-PASS RUMBLE FILTER—Second-order rumble filter for phonograph amplifier has 1-dB peak and 20-Hz cutoff frequency. Design uses large resistance values to permit use of smaller and lower-cost capacitors.—D. Lancaster, "Active-Filter Cookbook," Howard W. Sams, Indianapolis, IN, 1975, p 191–192.

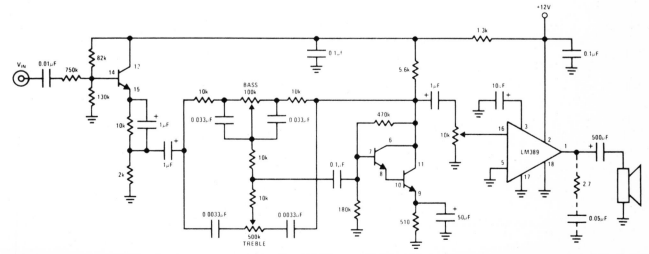

CERAMIC-CARTRIDGE SYSTEM—Circuit using National LM389 opamp having three transistors on same chip provides required high input impedance for ceramic cartridge because input transistor is wired as high-impedance emitter-follower. Remaining transistors form high-gain Darlington pair used as active element in low-distortion Baxandall tone-control circuit.—"Audio Handbook," National Semiconductor, Santa Clara, CA, 1977, p 4-33–4-37.

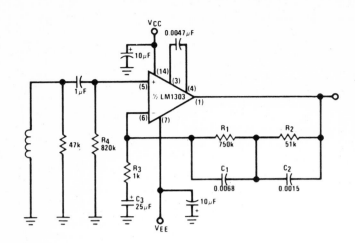

SPLIT-SUPPLY PHONO PREAMP—Low-noise circuit using LM1303 provides RIAA response and operates over supply voltage range of ±4.5 to ±15 V. 0-dB reference gain (1 kHz) is about 34 dB. Input is from magnetic cartridge.—"Audio Handbook," National Semiconductor, Santa Clara, CA, 1977, p 2-25–2-31.

SCRATCH/RUMBLE FILTER—Single active filter provides two widely differing turnover frequencies, as required in audio amplifier used with phonograph. For values shown, insertion loss of filter is −6 dB at 37 Hz and at 23 kHz. Components may be switched to provide different turnover frequencies, but complete removal of filter requires considerably more complicated switching.—P. I. Day, Combined Rumble and Scratch Filter, *Wireless World*, Dec. 1973, p 606.

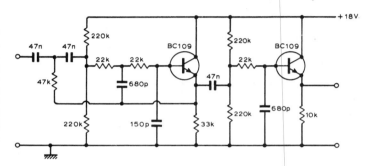

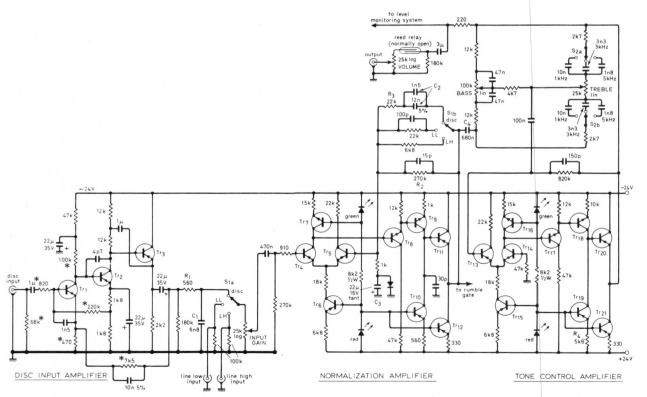

NO-COMPROMISE PHONO PREAMP—Distortion figure is below 0.002 percent, overload margin is about 47 dB, and S/N ratio is 71 dB for phono amplifier. This feeds normalization amplifier whose output is set at 0 dBm by setting input gain control. Feedback components R_2, R_3, and C_2 provide RIAA bass boost. Tone-control circuit is based on Baxandall system but has bass control turnover frequency which decreases as control approaches flat position. This allows small amount of boost at low end of audio spectrum to correct for transducer shortcomings. Article describes circuit operation in detail and gives additional circuits used for tape output, level detection, noise gate, and power supply. Transistors Tr_1-Tr_6 and Tr_{13}-Tr_{15} are BCY71; Tr_7-Tr_9 and Tr_{16}-Tr_{18} are MPS A06; Tr_{10}-Tr_{12} and Tr_{19}-Tr_{21} are MPS A56; Tr_9 is BFX85 or equivalent. Circuit is duplicated for other stereo channel.—D. Self, Advanced Preamplifier Design, *Wireless World*, Nov. 1976, p 41–46.

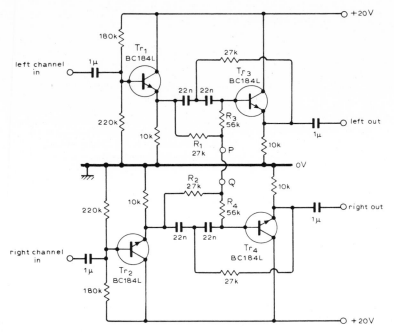

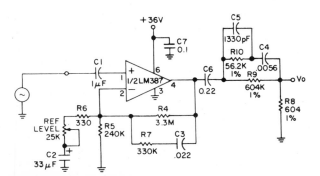

12-V PHONO PREAMP—Low-noise circuit has midband 0-dB reference gain of 46 dB. Designed for RIAA response. Internal resistor matrix of IC minimizes parts count. Input is from magnetic cartridge.—"Audio Handbook," National Semiconductor, Santa Clara, CA, 1977, p 2-25—2-31.

RUMBLE FILTER—Used when rumble from cheaper turntable or record extends above 100 Hz, causing disconcerting out-of-phase loudspeaker signals. Circuit is based on fact that human ear is not sensitive to directional information below about 400 Hz, making it permissible to remove stereo (L − R) signal at low frequencies and thus remove stereo rumble without losing stereo separation. Emitter-followers feed high-pass filters having 200-Hz breakpoint frequencies and Butterworth characteristics. Attenuation of filter is 12 dB at 100 Hz. Filter circuit can be disabled by placing switch between points P and Q.—M. L. Oldfield, Stereo Rumble Filter, *Wireless World,* Oct. 1975, p 474.

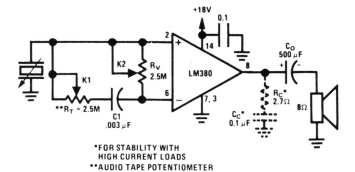

COMMON-MODE VOLUME AND TONE CONTROL—Eliminates attenuation of signal by conventional voltage-divider type of volume control and gives maximum input impedance. Used with transducers having high source impedance, but will also serve with low-impedance transducers.—"Audio Handbook," National Semiconductor, Santa Clara, CA, 1977, p 4-21—4-28.

INVERSE RIAA RESPONSE GENERATOR—Used in design, construction, and testing of phonograph preamp. Provides opposite of playback characteristic. Passive filter is added to output of National LM387, used as flat-response adjustable-gain block. Gain range is 24 to 60 dB, set in accordance with 0-dB reference gain (1 kHz) of preamp under test. Input is from 1-kHz square-wave generator, which can be built with other half of LM387 connected as also shown.—D. Bohn, Inverse RIAA/Square Wave Generator, *Audio,* Feb. 1977, p 65—66.

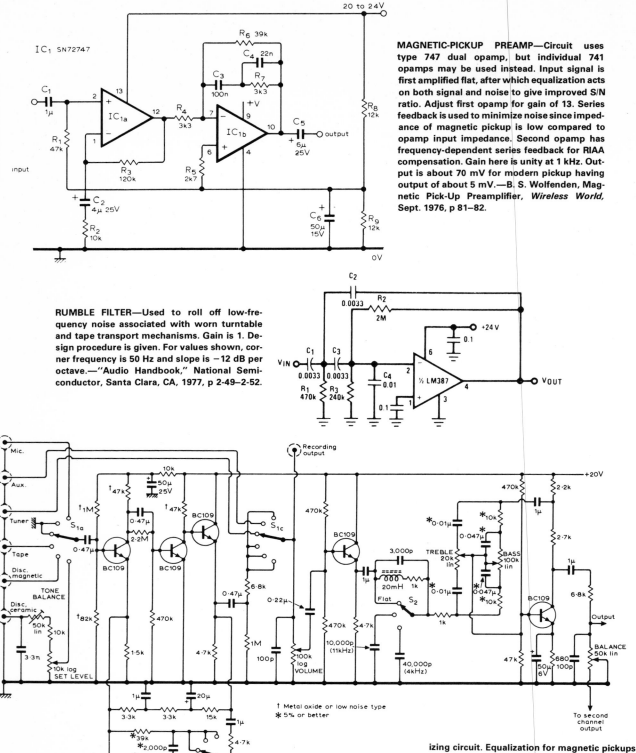

MAGNETIC-PICKUP PREAMP—Circuit uses type 747 dual opamp, but individual 741 opamps may be used instead. Input signal is first amplified flat, after which equalization acts on both signal and noise to give improved S/N ratio. Adjust first opamp for gain of 13. Series feedback is used to minimize noise since impedance of magnetic pickup is low compared to opamp input impedance. Second opamp has frequency-dependent series feedback for RIAA compensation. Gain here is unity at 1 kHz. Output is about 70 mV for modern pickup having output of about 5 mV.—B. S. Wolfenden, Magnetic Pick-Up Preamplifier, *Wireless World,* Sept. 1976, p 81–82.

RUMBLE FILTER—Used to roll off low-frequency noise associated with worn turntable and tape transport mechanisms. Gain is 1. Design procedure is given. For values shown, corner frequency is 50 Hz and slope is −12 dB per octave.—"Audio Handbook," National Semiconductor, Santa Clara, CA, 1977, p 2-49–2-52.

PREAMP WITH EQUALIZATION—Based on 1966 high-performance Bailey preamp design with improved filter and tone control circuits and additional complete ceramic-pickup equalizing circuit. Equalization for magnetic pickups and other types of inputs is automatically selected by three-deck input selector switch. To avoid overloading input stage, adjust set level control to give comfortable listening level for given input when main volume control is at about half its maximum rotation. Article also gives lower-cost version for ceramic-pickup equalization and changes required in this for operation from negative supply.—B. J. Burrows, Ceramic Pickup Equalization, *Wireless World,* Aug. 1971, p 379–382.

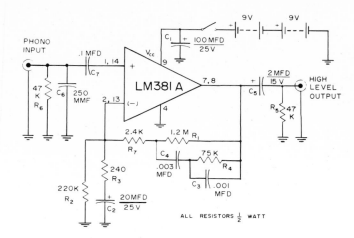

LOW-NOISE PREAMP—Provides dynamic range of about 80 dB for phonograph playback system, even when using highest-quality cartridge having low output. Source voltage is reduced to 18 V for National LM381A, which still provides ample signal for 2-V high-level input of stereo channel. Cross-channel isolation is better than 60 dB from 20 to 20,000 Hz.—J. P. Holm, A Quiet Phonograph Preamplifier, *Audio*, Oct. 1972, p 34–35.

CERAMIC-CARTRIDGE AMPLIFIER—Single National LM380 forms simple amplifier with tone and volume controls for driving 8-ohm loudspeaker at outputs above 3 W. Supply voltage range is 12–22 V, with higher voltage giving higher power. Tone control changes high-frequency rolloff.—"Audio Handbook," National Semiconductor, Santa Clara, CA, 1977, p 4-21–4-28.

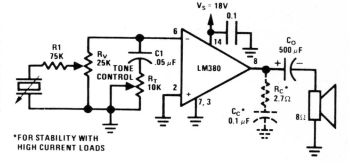

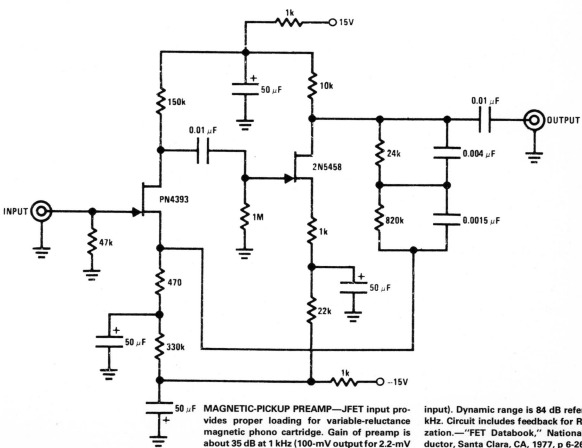

MAGNETIC-PICKUP PREAMP—JFET input provides proper loading for variable-reluctance magnetic phono cartridge. Gain of preamp is about 35 dB at 1 kHz (100-mV output for 2.2-mV input). Dynamic range is 84 dB referenced to 1 kHz. Circuit includes feedback for RIAA equalization.—"FET Databook," National Semiconductor, Santa Clara, CA, 1977, p 6-26–6-36.

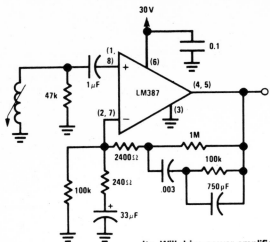

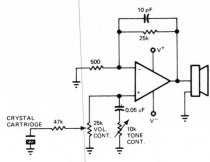

RIAA PHONO PREAMP—Design procedure is given for operation from 30-V supply, using magnetic cartridge having 0.5 mV/cm/s sensitiv- ity. Will drive power amplifier having 5 VRMS input overload limit.—"Audio Handbook," National Semiconductor, Santa Clara, CA, 1977, p 2-25–2-31.

5-W POWER OPAMP—Low-cost phono amplifier using only single 591 power opamp provides 5 W into 8-ohm load with only 0.2% total harmonic distortion. With crystal cartridge, circuit has fixed gain of 50.—R. J. Apfel, Power Op Amps—Their Innovative Circuits and Packaging Provide Designers with More Options, *EDN Magazine,* Sept. 5, 1977, p 141–144.

CHAPTER 14
Photography Circuits

Includes adjustable or programmable timers for enlargers and printers, photoflash, slave flash, strobe, and controlled-sequence flash circuits, exposure meters, and gray-scale control for CRT.

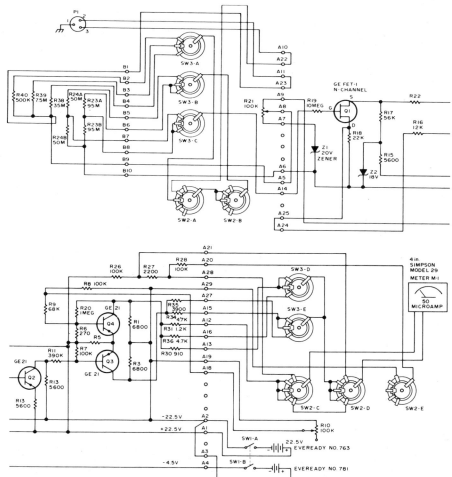

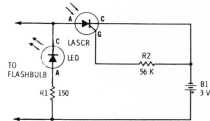

SLAVE FLASH—Remote flashtube having no connection with camera is fired by light-activated SCR (LASCR) when triggered by main flash of camera. Used to provide illumination at greater depth than main flash range, to soften sharp shadows, and to provide backlighting for flash photographs. LED is indicator showing that circuit has been triggered, reminding photographer that new flash lamp should be inserted.—F. M. Mims, "Transistor Projects, Vol. 1," Radio Shack, Fort Worth, TX, 1977, 2nd Ed., p 79–85.

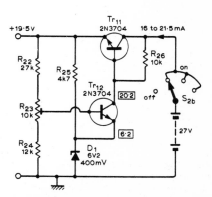

THREE-RANGE LIGHT METER—Uses probe containing Clairex 905HN light-dependent resistance element, connected to DC differential amplifier driving meter having specially calibrated scale. Article gives calibration procedure. Switching circuit provides constant check on voltage of 22.5-V battery. If 4.5-V battery is low, full-scale adjustment cannot be made. Resistors having values specified in article are connected in turn to terminals of photocell jack P1 for calibration that gives linear scale reading.—J. L. Mills, Jr., Light Right?—Do-It-Yourself Photo Exposure Meter, *73 Magazine*, Sept. 1978, p 204–206 and 208–211.

19.5 V FROM 27-V BATTERY—Used to provide precise voltage levels required by portable trigger unit designed to fire up to five different flash units at equal intervals that may range from 11 ms to 11 s. Article gives all circuits.—R. Lewis, Multi-Flash Trigger Unit, *Wireless World*, Nov. 1973, p 529–532.

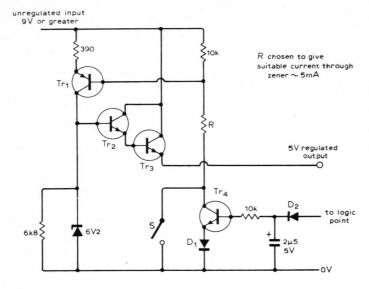

TIMER SWITCH-OFF FOR ENLARGER—Circuit shows power supply designed to operate digital exposure timer using TTL. Since timer logic is needed only when enlarger lamp is on, power supply circuit will be turned off automatically when timer goes low and turns off enlarger lamp at end of exposure. Switch S is closed when timer cycle is activated, setting timer output at 5 V and turning on lamp. Transistor and diode types are not critical. Since D_2 is connected to logic point of timer, power supply remains on when S is released, until completion of timer cycle.—E. R. Rumbo, Automatic Switch-Off Power Supply, *Wireless World,* Feb. 1976, p 77.

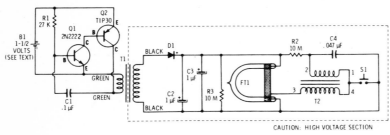

CAUTION: HIGH VOLTAGE SECTION

XENON STROBE—Two-transistor oscillator generates pulses at about 500 Hz for step-up by 300-mA filament transformer T1 (Radio Shack 273-1384) to charge storage capacitors C2 and C3, which are 250-V electrolytics. Simultaneously, C4 is charged through R2. After allowing sufficient time for capacitors to charge, S1 is pressed to discharge C4 through 272-1146 flashtube trigger transformer T2, which steps up voltage pulse to about 4000 V for ionizing gas in 272-1145 xenon flashtube FT1. C2 and C3 now discharge through ionized gas to produce brilliant flash of white light lasting only a few microseconds, as required for photography of objects moving at high speed. Circuit may require two cells in series for reliable operation.—F. M. Mims, "Transistor Projects, Vol. 3," Radio Shack, Fort Worth, TX, 1975, p 49–60.

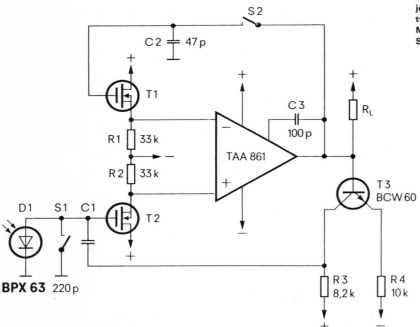

BPX 63 220 p

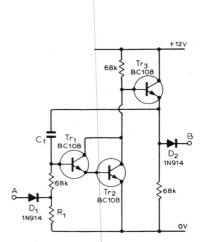

2-min RAMP—Used in multiple timer for development of photographic paper, in which six independent timers are started in sequence as each sheet of exposed paper is placed in developer. C_1 is 1 μF and R_1 is 11 megohms for 2-min timer having accuracy within 5 s. Article gives all other circuits required and suggests modifications to meet other needs. Output B drives meter and trigger circuit for audible alarm. Timer is started by input switch connected to A.—R. G. Wicker, Photographic Development Timer, *Wireless World,* April 1974, p 87–90.

LOW-LEVEL EXPOSURE METER—Uses Siemens BPX 63 photodiode having sensitivity of 10 nA per lux in circuit which ensures that aperture setting is affected only by useful light and not by noise signals. When used at low light levels, circuit recovers quickly from temporary light bursts. Switches S1 and S2 are closed when camera shutter is not open; opamp output is then connected to its inverting input through FET T1. At commencement of exposure, S1 and S2 open to give amplification of over 3000. Integrating capacitor C1 is then charged by photocurrent, making output voltage vary linearly with time. Base-emitter junction of T3 begins to conduct at output voltage of 1 V. Exposure is completed when C1 provides feedback via T3 so no current flows through load resistor R_L. Supply is ± 3 V.—"Photodiode BPX 63—All It Needs Is Starlight," Siemens, Iselin, NJ.

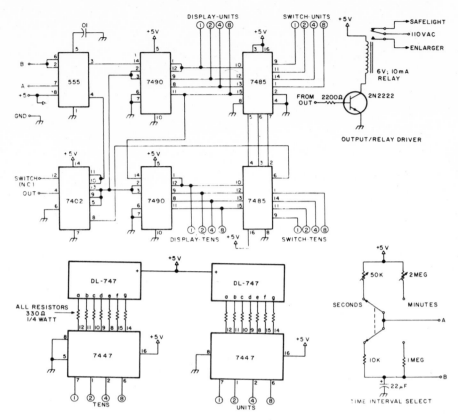

BCD THUMBWHEEL-SET 99-min TIMER—Provides timing in seconds to 99 s, and timing in minutes to 99 min, with 2-digit LED indicator showing elapsed time. Desired interval is set with BCD thumbwheel switches. LED readout counts up to preset time, then resets automatically to zero. Switch giving choice of seconds or minutes has center-off position that stops count temporarily for burning in portion of negative. Article gives construction details.—M. I. Leavey, Build a Unique Timer, *73 Magazine,* Aug. 1977, p 66–71.

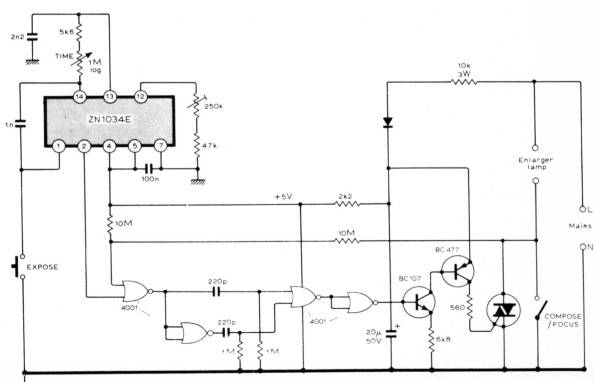

ENLARGER TIMER—Requires no transformer-type power supply because circuit operates from 1 mA taken from AC line through 10K resistor and rectifier. Ferranti ZN1034E timer IC generates delay and supplies 5 V for 4001 CMOS gates. Triac is triggered with 100-μs 60-mA pulses at zero-crossing point. Logarithmic time-control pot may be calibrated from 1 to 120 s. Choose triac to handle current drawn by enlarger lamp used.—M. J. Mayo, Transformerless Enlarger Timer, *Wireless World,* May 1978, p 68.

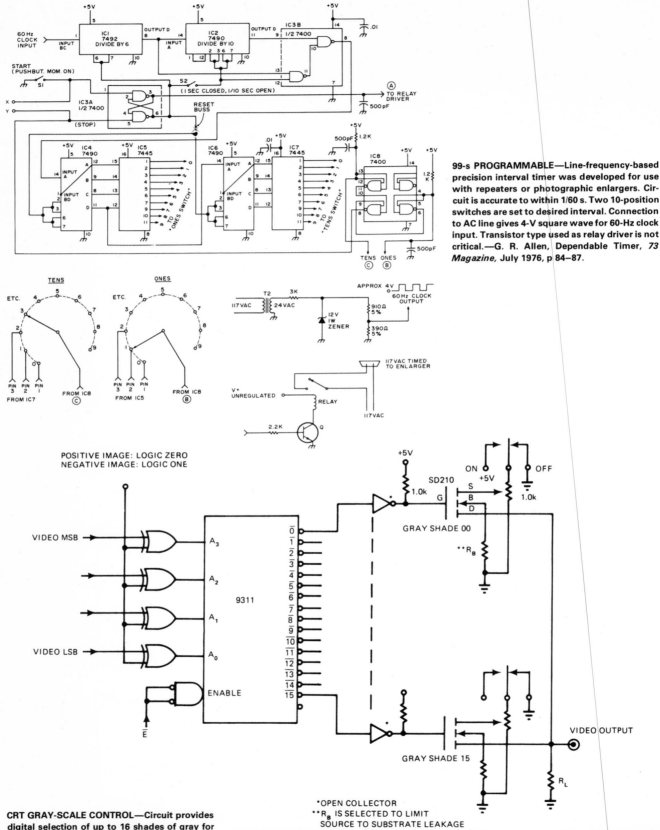

99-s PROGRAMMABLE—Line-frequency-based precision interval timer was developed for use with repeaters or photographic enlargers. Circuit is accurate to within 1/60 s. Two 10-position switches are set to desired interval. Connection to AC line gives 4-V square wave for 60-Hz clock input. Transistor type used as relay driver is not critical.—G. R. Allen, Dependable Timer, *73 Magazine,* July 1976, p 84—87.

CRT GRAY-SCALE CONTROL—Circuit provides digital selection of up to 16 shades of gray for image on screen of cathode-ray tube, as required for different imaging requirements or different photographic films. DMOS FETs provide fast switching times so data rate is limited only by TTL drive circuits. Four bits of digital data stored in 9311 memory are used for selecting desired scale. Output of circuit is used to control beam intensity. Circuit also permits complete video inversion for negative images.—K. R. Peterman, Fast CRT Intensity Selector Adjusts the Gray Scale, *EDN Magazine,* March 20, 1976, p 98 and 100.

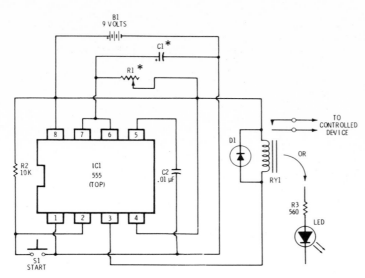

BASIC 555 TIMER—Closing switch S1 momentarily applies activating signal to trigger input pin 2 of timer, to start charging of C1. When C1 charges to two-thirds of supply voltage, timer discharges it to complete timing cycle. Duration of charging interval can be varied from several microseconds to over 5 min by changing values of R1 and C1. With 1K for R1, capacitor values of 0.01 to 100 μF give time range of 10 μs to 100 ms. With 100 megohms and 1 μF, time increases to 10 s. Once timer starts, closing S1 again has no effect. Timing cycle can be interrupted only by applying reset pulse to pin 4 or opening power supply. Circuit will drive LED directly or can be used with miniature relay (Radio Shack 275-004) to control larger loads. Can be used as darkroom timer if LED is kept several feet away from photographic paper. Diode is 1N914.—F. M. Mims, "Integrated Circuit Projects, Vol. 2," Radio Shack, Fort Worth, TX, 1977, 2nd Ed., p 57–65.

MULTIFLASH SWITCH—When ramp output of flash trigger circuit (given in article) is applied to input at A, flash at output of switch circuit is tripped when ramp voltage reaches level determined by setting of R_{12}. Similar voltage-operated switches are required for other flashes. Used for taking sequence photographs such as springboard diver in flight. Settings of R_{12} for different switches are chosen for equal times between flashes, with intervals from 11 ms to 11 s. Article gives all circuits and setup procedure. Regulated 19.5-V supply is required—R. Lewis, Multi-Flash Trigger Unit, *Wireless World*, Nov. 1973, p 529–532.

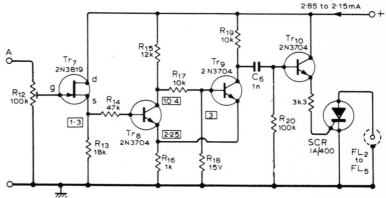

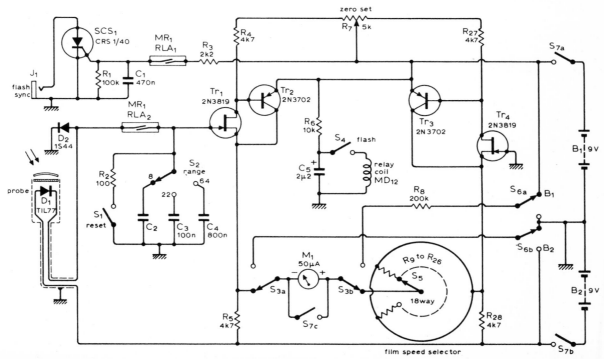

f-NUMBER FLASHMETER—Used to measure light produced at subject position by electronic flashlamps prior to actual taking of picture. Meter is calibrated to read correct f-number setting of lens aperture. Three ranges are provided, from f/2 to f/64, while film speed selector covers films from ASA 12 to 650. Texas Instruments TIL77 photodiode is used as sensing element in probe. Article covers construction, operation, and calibration of meter in detail. Table in article gives values for 18 resistors (one for each film speed) selected by S_5. Examples are 20K for ASA 64 and 51K for ASA 25.—R. Lewis, Photographic Flashmeter, *Wireless World*, Aug. 1974, p 273–278.

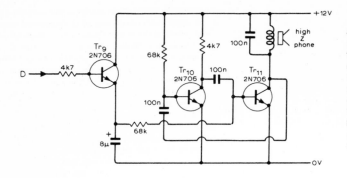

AUDIBLE ALARM FOR TIMER—Used with 2-min timer for developing photographic paper, to produce short warning bleep indicating end of developing time. Input D is taken from output of Schmitt trigger that changes state when 2-min ramp generator times out. Tr_9 and C_4 together lengthen short reset pulse so MVBR Tr_{10}-Tr_{11} oscillates long enough for signal to be heard.—R. G. Wicker, Photographic Development Timer, *Wireless World*, April 1974, p 87–90.

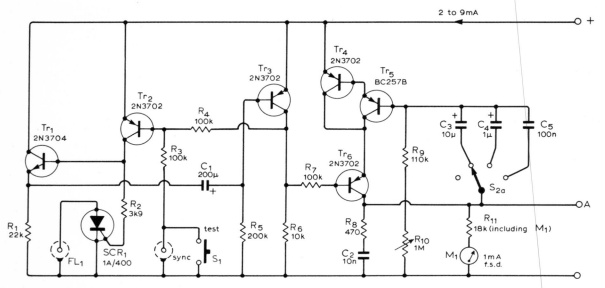

FLASH TRIGGER—Used in instrument designed to trigger up to five individual flash units at equal increments of time that can range from 11 ms to 11 s, as required for such assignments as taking sequence photographs of springboard diver in flight. Transistors Tr_1, Tr_2, and Tr_3 form monostable MVBR that is switched to unstable state by negative pulse applied to base of Tr_2 by SCR_1 when camera shutter contacts FL_1 are closed. Timing circuit Tr_4-Tr_5-Tr_6 provides ramp output at A for feeding voltage-operated switches set to trip at different points of ramp waveform as required for triggering flashes in sequence. Article gives all circuits and setup procedure. Regulated 19.5-V supply is required.—R. Lewis, Multi-Flash Trigger Unit, *Wireless World*, Nov. 1973, p 529–532.

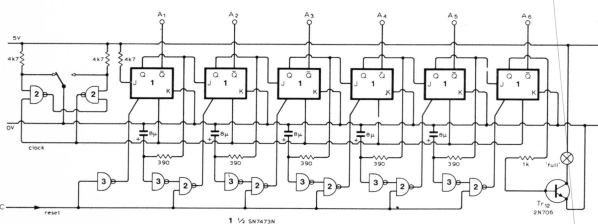

MULTIPLE TIMER FOR PRINTS—Six independent 2-min timers, each using half of SN7473N IC, are set in sequence by unique input switch as sheets of exposed paper are inserted in developer at about 20-s intervals. When capacity of six prints is reached, Tr_{12} turns on light to tell operator that no more prints should be inserted until control logic activates alarm signifying 2-min time for first sheet inserted. Audible bleep is repeated as each subsequent sheet reaches its 2-min development time. Article gives all circuits and explains operation in detail. Two-input NAND gates (each ¼ of SN7400N) and inverters (each ⅙ of SN7404N) are used to steer reset pulses. Similar two-input NAND gates are used to form fully compatible input pulses from input switch control, each having correct level, rise time, and fall time, without contact bounce that might cause spurious starting of several timers simultaneously.—R. G. Wicker, Photographic Development Timer, *Wireless World*, April 1974, p 87–90.

1 ½ SN7473N
2 ¼ SN7400N
3 ⅙ SN7404N

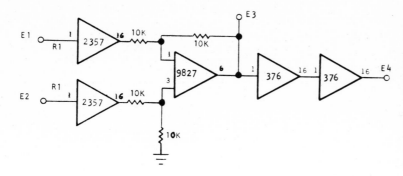

DENSITY AND EXPOSURE—Circuit converts transmission parameter of spectrophotometer to more useful density parameter, which in turn can be converted to exposure parameter. Optical Electronics 2357 opamps at input provide 90-dB dynamic range for DC to 1 kHz or 40-dB range for DC to 100 kHz, operating basically as current amplifiers. 9827 is used as wideband opamp in unity-gain subtracter configuration. Additional 376 opamps are used only for converting to exposure parameter. Use 1000 ohms for R1 with 10-V full-scale inputs.—"Conversion of Transmission to Density and Density to Exposure," Optical Electronics, Tucson, AZ, Application Tip 10133.

100-Ws PHOTOFLASH—Uses AC supply and large storage capacitors to give intense flash lasting only about 250 ms, as required for stop-motion photography of fast-moving objects such as bullets. For battery-powered operation, T1 can be replaced by solid-state chopper circuit. Contacts can be in camera or in external control device.—W. E. Hood, Lightning in a Bottle, *73 Magazine,* Sept. 1974, p 109–112.

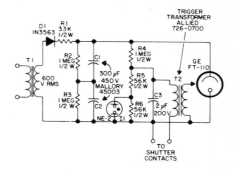

CHAPTER 15

Remote Control Circuits

Wired, wireless, light-beam, and other techniques are given for controlling transmitters, transceivers, receivers, motors, and other switched devices from a distance, including use of tone coders and decoders.

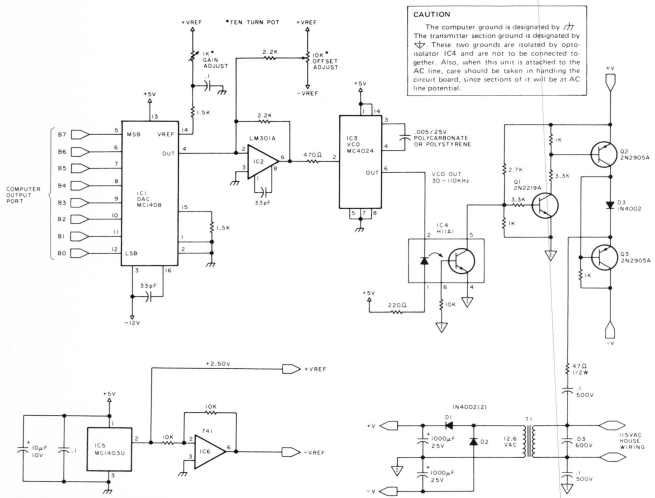

CARRIER-CURRENT TRANSMITTER—Modulates existing house wiring with high-frequency signals that can be detected by special receivers plugged into any AC outlet, for control of appliances by home computer. Applications include turning house lights on and off during owner's absence on elaborate time schedule programmed into computer. IC1 converts 8-bit data word from computer to proportional analog output current. This is converted to voltage by IC2 for control of VCO IC3 that gives frequency proportional to voltage. With values shown, range is about 30 to 110 kHz, with 256 discrete increments of frequency. Thus, input code 00000000 gives 30 kHz, 00000001 gives 30.3 kHz, and 01000000 (decimal 64) gives 49.2 kHz. Signal is applied to house wiring by 0.5-W power amplifier Q1-Q3, using optical coupling through IC4 to prevent computer circuit from interacting with house wiring. Supply voltage ±V is 11 to 13 V. T1 is 12.6-VAC 300-mA filament transformer. IC5 is 2 to 2.5 V reference chip such as MC1403U. System uses one frequency to turn receiver on and frequency 4 kHz above or below in 8-kHz band to turn it off, for maximum of ten control channels in system. Article covers calibration of transmitter.—S. Ciarcia, Tune in and Turn on, Part 1: A Computerized Wireless AC Control System, *BYTE,* April 1978, p 114–116, 118, 120, and 122–125 (Part 2—May 1978, p 97–100 and 102).

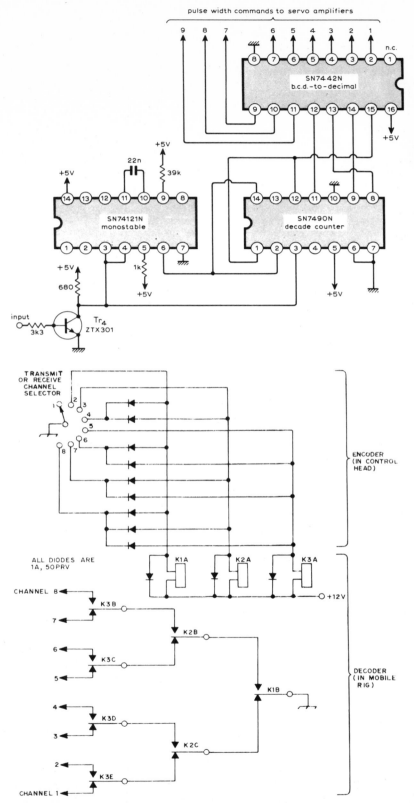

pulse width commands to servo amplifiers

NINE-CHANNEL DECODER—Circuit accepts serial information arriving over data link as series of nine varying-width pulses followed by fixed-width sync pulse, and after detection passes the nine individual commands to their respective servoamplifiers. Use of TTL ICs gives low component count for remote control system. Detection of sync pulse is done by comparing length of inverted input pulses with output of 0.6-ms monostable reference. All command pulses exceed 0.6 ms, so only 0.5-ms sync pulse clears counter to prepare for next channel-1 command pulse. Article gives operating details of system and circuits for coder and servoamplifier.—M. F. Bessant, Multi-Channel Proportional Remote Control, *Wireless World*, Oct. 1973, p 479–482.

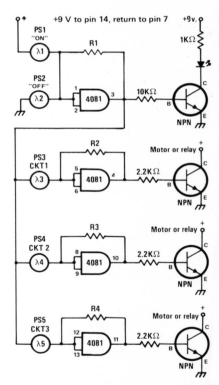

LIGHT BEAM FOR CONTROL OF MOVING TOY—Battery-powered CMOS logic is switched on and off by aiming flashlight beam at photocell, for turning small motors of model train or other powered toy on and off. Transistors can be 2N2222A for most small motors, but larger motors will require power transistors. Use high-intensity flashlight, with shield over lens to restrict beam width, so only one of five photocells is illuminated at a time. LED shows ON/OFF status of circuit. Values of R1-R4 are chosen so each gate flips logic state only when associated photocell is illuminated.—J. Sandler, 9 Projects under $9, *Modern Electronics*, Sept. 1978, p 35–39.

8 CHOICES WITH 3 WIRES—Provides remotely selected choice of eight functions, such as channels in mobile FM station, with only three wires running from control head to controlled equipment that can be in front of car. System involves converting 8-position switch selection in control head to 3-bit binary form for three control wires going to three-relay arrangement for decoding back to 8-position format. Relays are two-pole and four-pole double-throw 12-V units.—G. D. Rose, Independent 8-Channel Frequency Selection with Only Three Wires, *QST*, Aug. 1974, p 36–40.

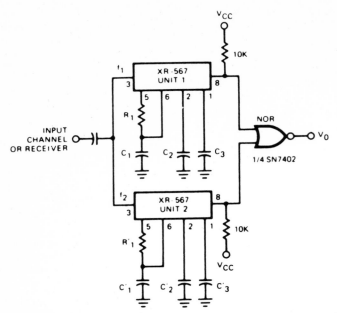

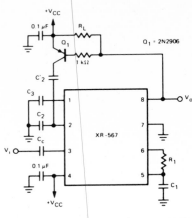

DUAL TONE DECODER—Used in communication systems where control or other information is transmitted as two simultaneous but separate tones. Circuit uses two Exar XR-567 PLL units in parallel, with resistor and capacitor values of each PLL decoder selected to provide desired center frequencies and bandwidth requirements. Supply voltage is 5–9 V.—"Phase-Locked Loop Data Book," Exar Integrated Systems, Sunnyvale, CA, 1978, p 41–48.

DUAL TIME-CONSTANT TONE DECODER—Exar XR-567 PLL system is connected as decoder having narrow bandwidth and fast response time. Circuit has two low-pass loop filter capacitors, C_2 and C'_2. With no input, pin 8 is high, Q_1 is off, and C'_2 is out of circuit. Filter then has only C_2, which is kept small for minimum response time. When in-band input tone signal is detected, pin 8 goes low, Q_1 turns on, and C'_2 is in parallel with C_2 to give narrow bandwidth. Supply voltage can be 5–9 V.—"Phase-Locked Loop Data Book," Exar Integrated Systems, Sunnyvale, CA, 1978, p 41–48.

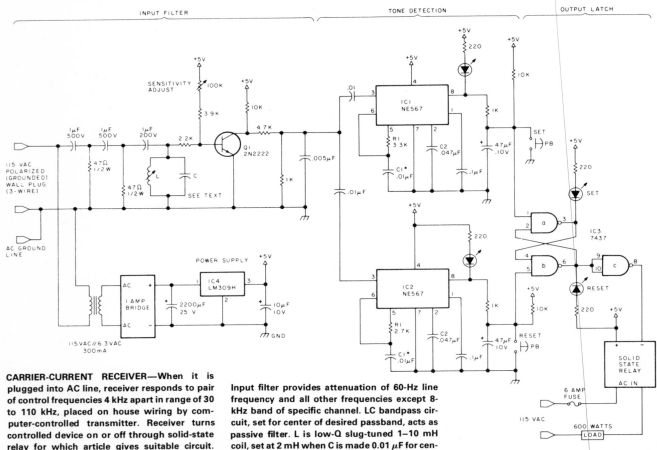

CARRIER-CURRENT RECEIVER—When it is plugged into AC line, receiver responds to pair of control frequencies 4 kHz apart in range of 30 to 110 kHz, placed on house wiring by computer-controlled transmitter. Receiver turns controlled device on or off through solid-state relay for which article gives suitable circuit. Tuned bandpass filter amplifies only that pair of frequencies assigned to its receiver, attenuating all other frequency pairs used in system. Amplified signal is sent to tone decoders IC1 and IC2, one responding to each frequency.

Input filter provides attenuation of 60-Hz line frequency and all other frequencies except 8-kHz band of specific channel. LC bandpass circuit, set for center of desired passband, acts as passive filter. L is low-Q slug-tuned 1–10 mH coil, set at 2 mH when C is made 0.01 µF for center frequency of 35 kHz. Article covers operation in detail and gives procedure for determining values of R1, C1, and C2 for each detector. Solid-state output relay can be Sigma 226 RE1-5A1, rated 6 A.—S. Ciarcia, Tune in and Turn on, Part

2: An AC Wireless Remote Control System, *BYTE*, May 1978, p 97–100 and 102 (Part 1—April 1978, p 114–116, 118, 120, and 122–125).

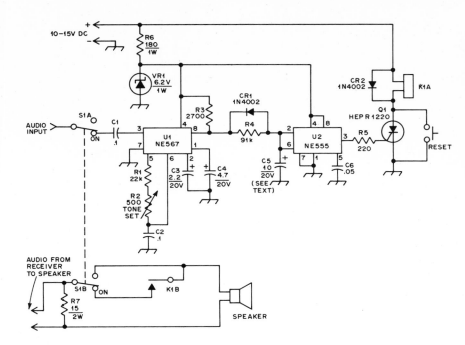

TIMED-TONE DECODER—Uses NE567 PLL and 555 timer to activate muted monitoring receiver until alerting audio tone of correct frequency and duration is received. Can be applied to almost any receiver for weather emergency alert warnings, paging calls, and similar services without having to listen continously to other traffic on channel. If received tone is within bandwidth of tone decoder, output of U1 goes nearly to zero and C5 starts to discharge through R4. When voltage at pins 2 and 6 of U2 reaches one-third of supply voltage, output of U2 goes high and triggers SCR Q1, energizing 12-V relay K1. Values shown for C4 and C5 give 1-s delay, which means triggering tone must be on at least 1 s. Once SCR is triggered, it holds relay on even after tone ceases. Pushbutton switch shorts SCR and releases relay when reset is desired. Zener provides regulated 6.2 V required for decoder. Values shown for R1, R2, and C2 give response to 450-Hz tone. Avoid use of Touch-Tone frequency, to prevent accidental triggering by those using Touch-Tone system.—J. S. Paquette, A Time-Delayed Tone Decoder, *QST,* Feb. 1977, p 16–17.

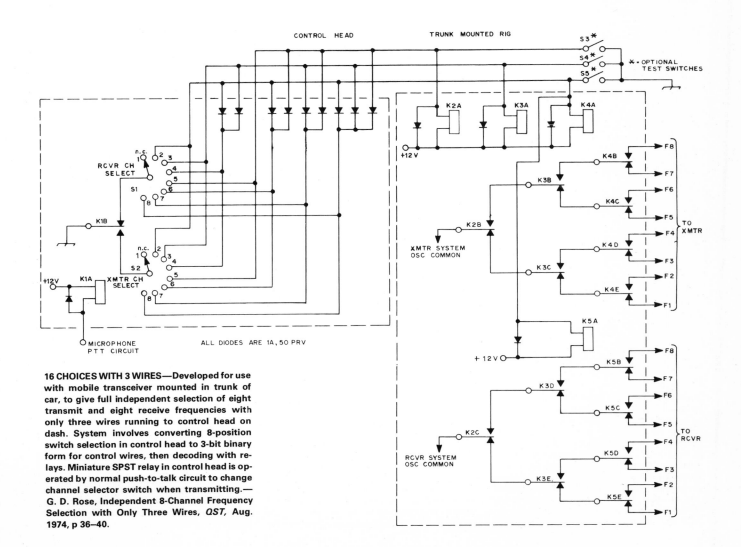

16 CHOICES WITH 3 WIRES—Developed for use with mobile transceiver mounted in trunk of car, to give full independent selection of eight transmit and eight receive frequencies with only three wires running to control head on dash. System involves converting 8-position switch selection in control head to 3-bit binary form for control wires, then decoding with relays. Miniature SPST relay in control head is operated by normal push-to-talk circuit to change channel selector switch when transmitting.— G. D. Rose, Independent 8-Channel Frequency Selection with Only Three Wires, *QST,* Aug. 1974, p 36–40.

REMOTE SWITCHING—Uses four flip-flops, each having one 4-input and one 2-input CMOS NAND gate. Momentarily grounding any input drives corresponding output high and all other outputs low. Unless power is interrupted, additional pulses on same input have no effect; circuit remains stable until some other input is momentarily grounded. Outputs can be used to drive other logic devices directly or through buffer if current required exceeds 10 mA. Can be used for remote frequency control of VHF transceiver and for other applications requiring remote selection of mutually exclusive functions.—P. Shreve, Remote-Switching Circuit, *Ham Radio*, March 1978, p 114.

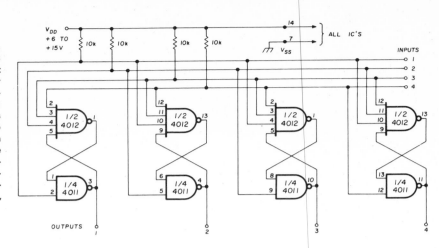

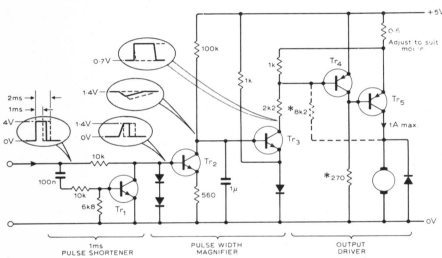

RADIO CONTROL FOR MOTOR—Proportional control system produces control pulses every 20 ms, with length of each adjustable between 1 and 2 ms. Circuit removes first 1 ms of pulse and expands remainder to produce 0–20 ms pulses for driving motor. Pulsing of motor gives smoother control than resistors, particularly at very low speeds. Transistor types are not critical. Tr_5 can be OC28. Optional dashed connection of 8.2K resistor provides foldback current/voltage protection.—M. Weston, Variable-Speed Radio Control Motor, *Wireless World*, Feb. 1978, p 59.

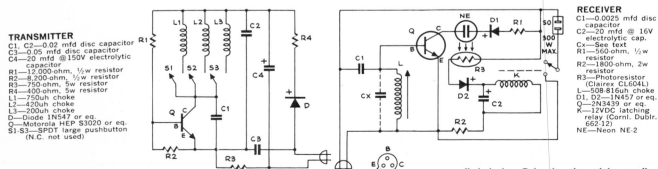

TRANSMITTER
C1, C2—0.02 mfd disc capacitor
C3—0.05 mfd disc capacitor
C4—20 mfd @150V electrolytic capacitor
R1—12,000-ohm, ½w resistor
R2—8,200-ohm, ½w resistor
R3—750-ohm, 5w resistor
R4—400-ohm, 5w resistor
L1—750uh choke
L2—420uh choke
L3—200uh choke
D—Diode 1N547 or eq.
Q—Motorola HEP S3020 or eq.
S1-S3—SPDT large pushbutton (N.C. not used)

RECEIVER
C1—0.0025 mfd disc capacitor
C2—20 mfd @ 16V electrolytic cap.
Cx—See text
R1—560-ohm, ½w resistor
R2—1800-ohm, 2w resistor
R3—Photoresistor (Clairex CL604L)
L—508-816uh choke
D1, D2—1N457 or eq.
Q—2N3439 or eq.
K—12VDC latching relay (Cornl. Dublr. 662-12)
NE—Neon NE-2

WIRELESS CONTROL—Choice of three lamps or appliances anywhere in house and garage, or even in neighboring home if on the same power transformer, can be turned on or off individually with three-channel transmitter that plugs into any wall outlet. Transmitter injects one of three tones (depending on button pushed) into house wiring. Receivers at locations of controlled devices are each tuned to one of carrier tones. Correct tone for receiver energizes neon lamp, and resulting light is picked up by photoresistor that energizes latching relay K for turning on controlled device. Relay is released by sending same tone again. Values of CX can be 0.005, 0.01, and 0.02 μF. Adjust slug of L2 for each receiver so neon comes on when assigned tone for that receiver arrives.—W. J. Hawkins, Three-Channel Wireless Switch—Use It Anywhere, *Popular Science*, Sept. 1973, p 98–99 and 121.

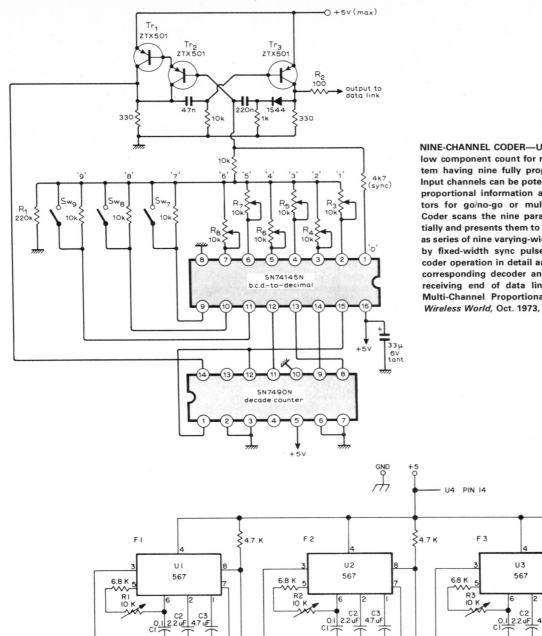

NINE-CHANNEL CODER—Use of TTL ICs gives low component count for remote control system having nine fully proportional channels. Input channels can be potentiometers for fully proportional information and switched resistors for go/no-go or multistep information. Coder scans the nine parallel inputs sequentially and presents them to single-line data link as series of nine varying-width pulses followed by fixed-width sync pulse. Article describes coder operation in detail and gives circuits for corresponding decoder and servoamplifier at receiving end of data link.—M. F. Bessant, Multi-Channel Proportional Remote Control, *Wireless World,* Oct. 1973, p 479–482.

ON/OFF CONTROL BY THREE TONES—Used for decoding two Touch-Tone digits to give operation or release of relay by remote control over wire line. Three 567 tone decoders and 7402 quad gate are adjusted to recognize tones corresponding to any two keys in given row or column on Touch-Tone keyboard. As example, * key generates 941 and 1209 Hz, and circuit can be adjusted so these two frequencies energize relay. Similarly, pushing of # key generates 941 and 1477 Hz that can be used for deenergizing relay.—W. J. Hosking, Simple New TT Decoder, *73 Magazine,* April 1976, p 52–53.

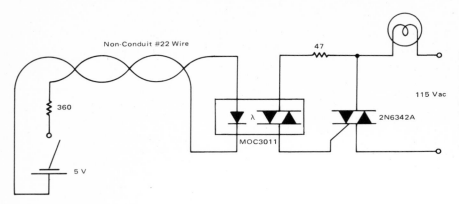

LAMP CONTROL WITHOUT CONDUIT—Motorola MOC3011 optoisolator permits control of large lamp, motor, pool pump, and other AC loads from remote location with low-voltage signal wiring while meeting building codes. Choice of triac depends on load being handled.—P. O'Neil, "Applications of the MOC3011 Triac Driver," Motorola, Phoenix, AZ, 1978, AN-780, p 5.

WIRELESS REMOTE TUNING—Frequency-to-voltage converter for transceiver responds to AF output of control receiver and feeds corresponding DC voltage to varactor tuning diode in VFO of transceiver, for remote wireless tuning. In most cases only a few volts of DC variation across varactor are sufficient, so variable audio oscillator at remote-control location need have range of only a few kilohertz.—J. Schultz, H.F. Operating—Remote Control Style, *CQ,* March 1978, p 22–23 and 90.

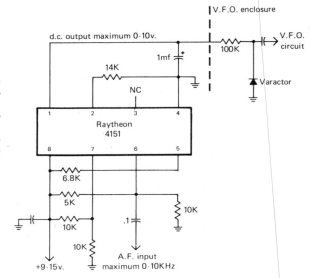

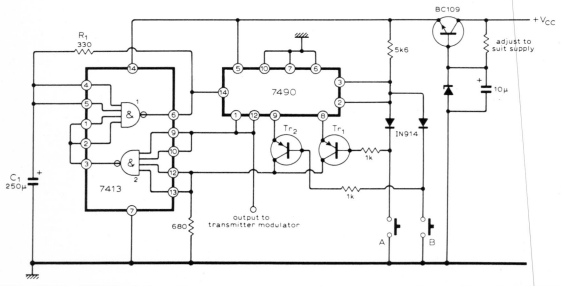

BLIP-AND-HOLD RADIO CONTROL—Coder uses two ICs to generate sequence of pulses suitable for actuators of radio control system. During standby, oscillator formed by NAND gate 1 operates at 0.5 Hz as determined by C_1 and R_1, and all four outputs of 7490 IC are zero. When switch A is closed, 7490 is clocked by negative edge of oscillator waveform and Tr_1 becomes forward-biased. Output of NAND gate 2 then drops to zero, stopping oscillator and holding outputs of 7490. When switch A is opened, outputs of 7490 again drop to zero. Many different blip-and-hold combinations can be obtained by suitable arrangement of switches and gates.—G. D. Southern, Sequence Generator for Radio Control, *Wireless World,* Jan. 1976, p 60.

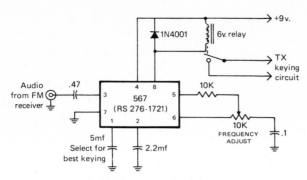

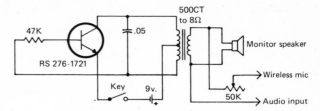

PLL TONE DECODER—Used in simple wireless FM remote control set up for keying transmitter. Keyed 500-Hz tone output of FM receiver at transmitter site acts through 567 PLL to operate 6-V relay whose contacts are in keying circuit of transmitter.—J. Schultz, H.F. Operating—Remote Control Style, *CQ*, March 1978, p 22–23 and 90.

500-Hz CONTROL TONE—Developed for use as wireless FM remote control for keying transmitter at another location by sending keyed audio tone over radio link, acoustic link fed by loudspeaker, or audio line. Frequency is about 500 Hz.—J. Schultz, H.F. Operating—Remote Control Style, *CQ*, March 1978, p 22–23 and 90.

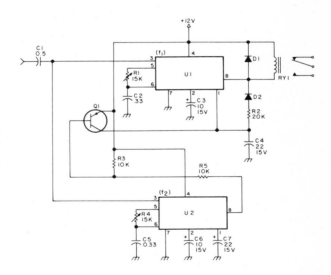

TWO-TONE CONTROL—Used to perform simple ON/OFF auxiliary function via repeater input. Two 567 decoders energize relay for input tone of 1800 Hz, with latching, and release it for 1950 Hz. Diodes are 1N4001. Relay can be 12 or 24 V. Q1 is 2N3905, 2N3906, MPS6521, or 2N2222.—W. Hosking, A Single Tone Can Do It, *73 Magazine*, Nov. 1977, p 184–185.

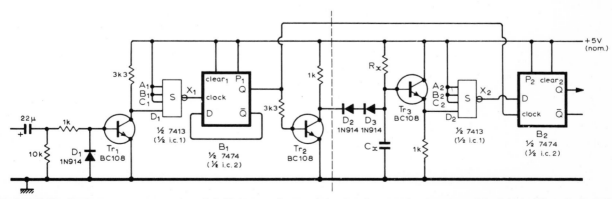

TONE DECODER—Replaces resonant reeds commonly used in multichannel radio-controlled models to detect modulation frequency being transmitted. Use of IC logic has advantage that range of audio frequencies can exceed an octave, whereas reeds cannot because they respond to second harmonic. Decoder has digital high-pass characteristic that is passed through inverter to give digital low-pass characteristic. Values of R_x and C_x determine critical frequency; for 900 Hz, use 150,000 ohms and 0.015 μF. To obtain n nonoverlapping bandpass characteristics, $n - 1$ basic elements with different critical frequencies are required; components to left of dashed line may be common to all these elements. Article covers multichannel systems in detail, along with use of time-division multiplexing.—C. Attenborough, Radio Control Tone Decoder, *Wireless World,* Dec. 1973, p 593–594.

CHAPTER 16
Siren Circuits

Includes variety of circuits for simulating sounds of police and other emergency sirens. Battery-operated versions can be used in toys or as part of burglar or fire alarm system. Some have adjustments for frequency, whooping rate, and duration of rising and falling tones.

10-W AUTO ALARM SIREN—Generates force field of high-intensity sound inside car, painful enough to discourage thief from entering car after tripping alarm switch by opening door. Circuit produces square-wave output that sweeps up and down in frequency. Modulation is provided by triangle waveform generated by R1, D1, and C1. If sweep-frequency siren is prohibited, remove C1 to produce legal two-tone sound. Use efficient horn loudspeaker capable of handling up to 10 W. D2 is silicon rectifier rated 1 A at 50 PIV. Other diodes are general-purpose silicon.—A. T. Roderick III, New Protection for Your Car, *73 Magazine,* March 1978, p 76–77.

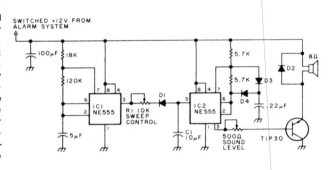

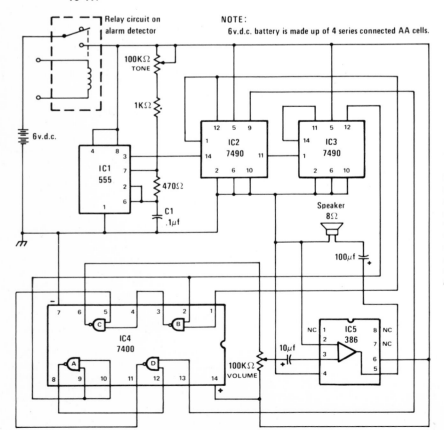

Relay circuit on alarm detector

NOTE:
6v.d.c. battery is made up of 4 series connected AA cells.

LOW-NOTE SIREN—Produces up/down blooping sounds characteristic of European police cars and now being used on some US emergency vehicles. Can be connected to burglar or theft alarm system for protection purposes, or used as portable sound box operated by momentary pushbutton switch. Includes volume control and tone control that varies both pitch and rate.—D. Heiserman, Whizbox, *Modern Electronics,* June 1978, p 67.

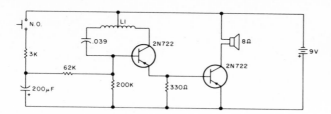

SIREN—Creates sounds resembling those of police-car siren in which air is forced through slots in motor-driven disk. L1 is half of audio transformer, using winding having 10K center tap.—Circuits, *73 Magazine*, April 1977, p 164.

FIRE SIREN USES FLASHER—Low-drain circuit operating from 1.5-V cell uses National LM3909 flasher IC to simulate fire-alarm siren. Pressing button produces rapidly rising wail, with tone coasting down in frequency after button is released. Sound from loudspeaker resembles that of motor-driven siren. Volume is adequate for child's pedal car.—P. Lefferts, Power-Miser Flasher IC Has Many Novel Applications, *EDN Magazine*, March 20, 1976, p 59—66.

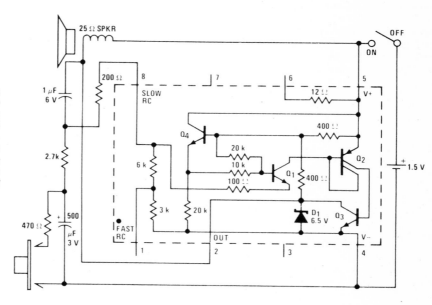

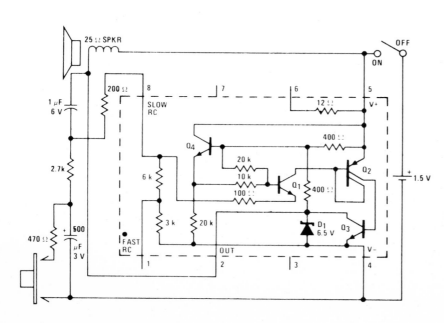

FIRE SIREN—Pressing button produces rapidly rising wail, and releasing button gives slower lowering of frequency resembling sounds of typical siren on fire engine. Circuit uses National LM3909 IC operating from 1.5-V cell for driving 25-ohm loudspeaker. 1-μF capacitor and 200-ohm resistor determine width of loudspeaker pulse, while 2.7K resistor and 500-μF capacitor determine repetition rate of pulses.—"Linear Applications, Vol. 2," National Semiconductor, Santa Clara, CA, 1976, AN-154, p 6—7.

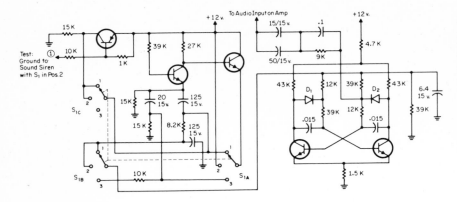

POLICE SIREN—Circuit used in Dietz siren-light police-car system gives distinctive tones. Position 1 of S_1 produces slow continuous rise and fall. Position 3 produces fast rising and falling tone. Position 2 rises slowly to full pitch when point 1 is grounded, then decays at same rate when point 1 is ungrounded. Position 3 gives most noticeable tone for break-in alarm on car. Terminal 1 goes to normally open door, hood, and other switches that complete circuit to ground when opened by intruder. Audio transistors and diode are general replacement types.—J. W. Crawford, The Ultimate Auto Alarm—Model II, *CQ*, Aug. 1971, p 54–57 and 96.

VARIABLE FREQUENCY AND RATE—Uses National LM380 opamp as astable oscillator with frequency determined by R_2 and C_2. Base of Q_1 is driven by output of LM3900 opamp connected as second astable oscillator, to turn output of LM380 on and off at rate fixed by R_1 and C_1. Transistor type is not critical. Circuit is ideal for experimenters.—"Audio Handbook," National Semiconductor, Santa Clara, CA, 1977, p 4-21–4-28.

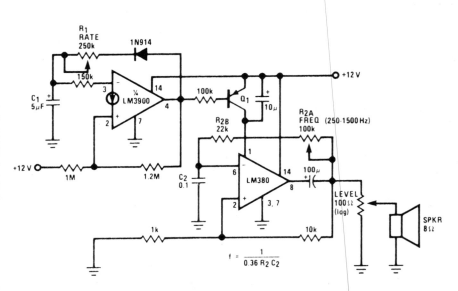

$$f = \frac{1}{0.36\,R_2\,C_2}$$

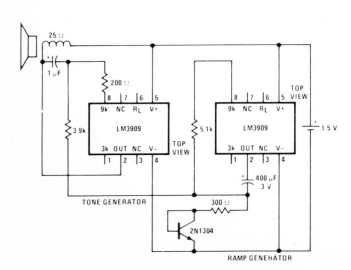

WHOOPER—Two National LM3909 ICs and single transistor generate rapidly modulated tone resembling that used on some police cars, ambulances, and airport emergency vehicles. Rapidly rising and falling modulating voltage is generated by IC having 400-μF capacitor. Diode-connected transistor forces this IC ramp generator to have longer ON periods than OFF periods, raising average tone of tone generator and making modulations seem more even.—"Linear Applications, Vol. 2," National Semiconductor, Santa Clara, CA, 1976, AN-154, p 7.

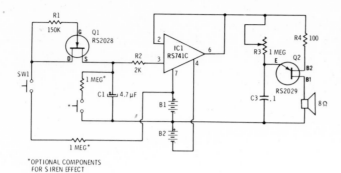

VARIABLE TONE USING VCO—Tone generator uses UJT and opamp in voltage-controlled oscillator. Frequency of audio output is determined by setting of R3. For two-tone siren effects, optional switches and resistors can be used. To speed up siren effect, use smaller value for C1.—F. M. Mims, "Integrated Circuit Projects, Vol. 4," Radio Shack, Fort Worth, TX, 1977, 2nd Ed., p 61–69.

*OPTIONAL COMPONENTS FOR SIREN EFFECT

LOUD BIKE SIREN—Uses 5558 dual opamp and four general-purpose NPN transistors to generate triangle wave that can be distorted by 10K symmetry control to give either fast or slow rise for sawtooth applied as base bias to astable MVBR Q1-Q2. Drain is reasonably low with 9-V radio battery. Repetition rate can be varied from long wail to rapid warble, and volume changed from soft to annoying. Article gives construction details, and recommends use of removable mounting on bike to avoid theft.—R. Megirian, Simple Electronic Siren, *73 Magazine*, Oct. 1977, p 176–177.

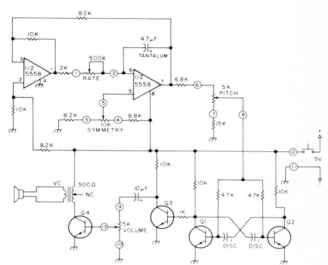

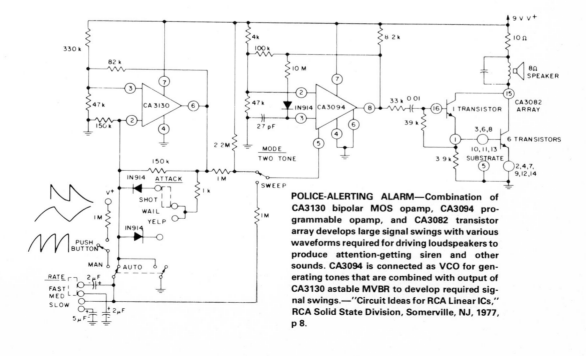

POLICE-ALERTING ALARM—Combination of CA3130 bipolar MOS opamp, CA3094 programmable opamp, and CA3082 transistor array develops large signal swings with various waveforms required for driving loudspeakers to produce attention-getting siren and other sounds. CA3094 is connected as VCO for generating tones that are combined with output of CA3130 astable MVBR to develop required signal swings.—"Circuit Ideas for RCA Linear ICs," RCA Solid State Division, Somerville, NJ, 1977, p 8.

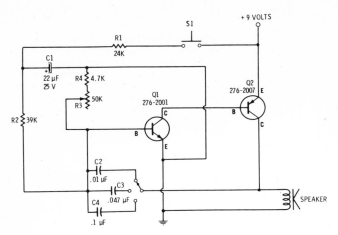

ADJUSTABLE SIREN—Tone is made adjustable by using multiposition switch to change capacitors in oscillator circuit. Speed (rate of change in frequency) of siren is adjusted with R3. 4700-ohm resistor in series with R3 keeps siren operational when R3 is rotated to minimum-resistance position. Siren is operated by pressing switch to produce rising wail, then releasing switch until wail drops down to cutoff.—F. M. Mims, "Transistor Projects, Vol. 1," Radio Shack, Fort Worth, TX, 1977, 2nd Ed., p 58—63.

POLICE SIREN USES FLASHER—Low-drain circuit operating from 1.5-V cell uses National LM3909 flasher ICs to simulate "whooper" sounds of electronic sirens used on some city police cars and ambulances. Two flashers are required for generating required rapidly rising and falling modulating voltage. Transistor is connected as diode to force ramp generator of IC to have longer ON periods than OFF periods, raising average tone and making modulation seem more even.—P. Lefferts, Power-Miser Flasher IC Has Many Novel Applications, *EDN Magazine,* March 20, 1976, p 59—66.

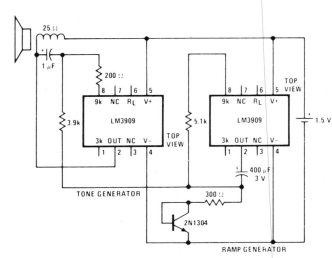

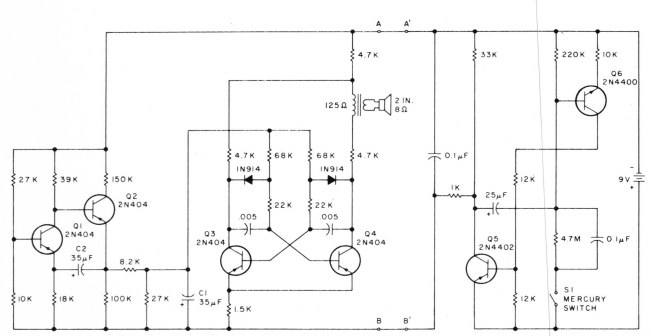

PORTABLE TOY SIREN—Can be assembled in small box as toy for small child. If mercury switch is used for S1, siren comes on automatically when box is picked up. MVBR Q1-Q2 controls rate at which siren wails, while Q3 and Q4 form AF MVBR that produces actual siren sound with frequency varied by triangle waveform on C1. MVBR Q5-Q6 is mono that conducts for preset time period when S1 is closed, for applying power to siren. Values shown give 12 s of operation before siren is shut off. When carried by child, siren is jostled enough so it keeps recycling.—J. H. Everhart, Super Siren, *73 Magazine,* Feb. 1978, p 96—97.

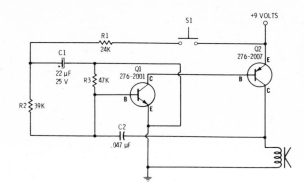

MANUALLY CONTROLLED SIREN—When switch is pressed, output tone of loudspeaker builds from low to high frequency. Releasing switch brings high frequency slowly back to low point and then cutoff. Siren sounds can be varied manually by pushing and releasing switch at different points in cycle. C2 controls pitch, and R3 determines speed at which pitch changes.—F. M. Mims, "Transistor Projects, Vol. 1," Radio Shack, Fort Worth, TX, 1977, 2nd Ed., p 58–63.

SIREN WITH MUTING—National LM389 array having three transistors and power opamp on same chip uses opamp as square-wave oscillator whose frequency is adjusted with R2B. One transistor is used in muting circuit to gate power amplifier on and off, while other two transistors form cross-coupled MVBR that controls rate of square-wave oscillator.—"Audio Handbook," National Semiconductor, Santa Clara, CA, 1977, p 4-33–4-37.

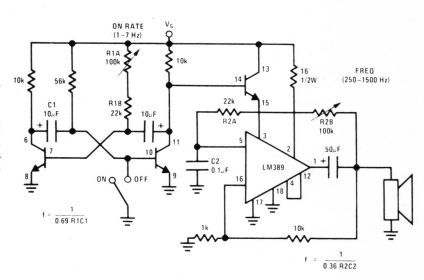

$$f = \frac{1}{0.69\,R1C1}$$

$$f = \frac{1}{0.36\,R2C2}$$

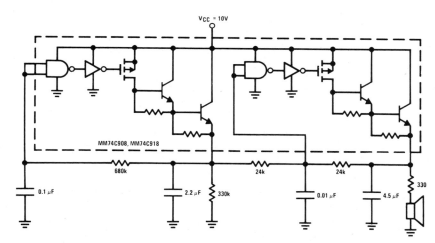

10-V SIREN CHIP—One section of National MM74C908/MM74C918 dual CMOS driver is used as audio VCO and other section as voltage ramp generator that varies frequency of VCO. Combination gives siren effect at low cost, with output current up to 250 mA for driving loudspeaker.—"CMOS Databook," National Semiconductor, Santa Clara, CA, 1977, p 5-38–5-49.

CHAPTER 17
Stereo Circuits

Includes amplifier and signal-processing circuits developed specifically for stereo FM, tape recorder, and phonograph systems. Many can be used singly in monophonic systems. Includes circuits for FM noise suppression, reverberation, rear-channel ambience, and loudspeaker phasing.

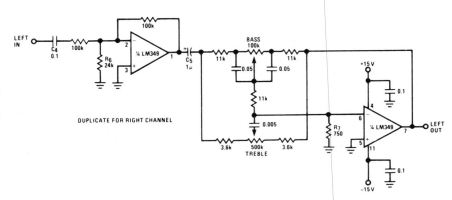

DUPLICATE FOR RIGHT CHANNEL

ACTIVE TONE CONTROLS—Provides ±20 dB gain with 3-dB corners at 30 and 10,000 Hz. Use of LM349 quad opamp means only one IC is needed for both stereo channels. Buffer at input gives high input impedance (100K) for source. Total harmonic distortion is typically 0.05% across audio band. Input-to-output gain is at least 5.—"Audio Handbook," National Semiconductor, Santa Clara, CA, 1977, p 2-40–2-49.

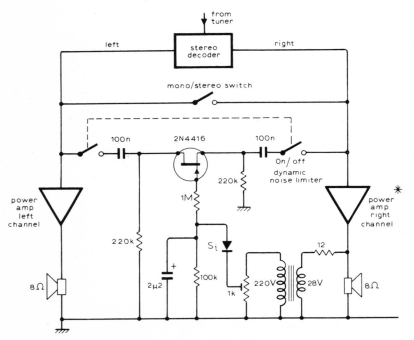

(✳ includes; volume, balance and tone controls)

FM NOISE SUPPRESSOR—Circuit acts as noise limiter to help produce pseudostereo sound having reduced noise, to offset noise signal heard during weak passages during stereo reception of FM stations. FET short-circuits both audio channels when audio signal strength drops sufficiently to make noise objectionable. If this voltage is insufficient to drive FET, amplifier or transformer must be used.—J. W. Richter, Stereo Dynamic Noise Limiter, *Wireless World,* Oct. 1975, p 474.

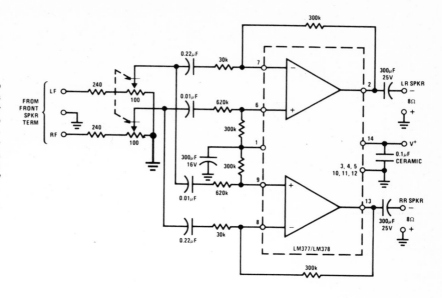

REAR-CHANNEL AMBIENCE—Can be added to existing left front and right front loudspeakers of stereo system to extract difference signal for combining with some direct signal (R or L) to add fullness for concert-hall realism during reproduction of recorded music. Very little power is required for pair of rear loudspeakers, and this can be furnished by National LM377/LM378 dual-amplifier IC operating from about 24-V supply.—"Audio Handbook," National Semiconductor, Santa Clara, CA, 1977, p 4-8–4-20.

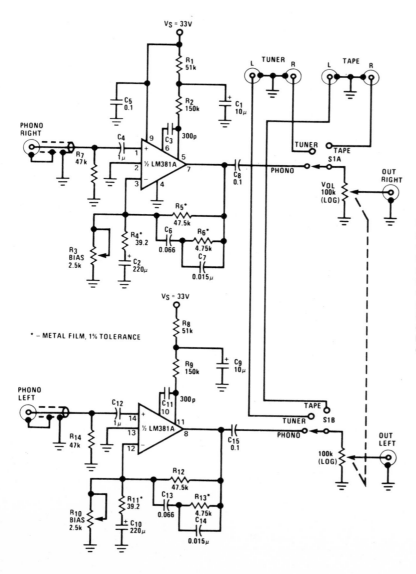

ULTRALOW-NOISE PREAMP—Complete preamp has inputs for magnetic-cartridge pickup, tuner, and tape, along with ganged volume control and ganged selector switch for both channels. Tone controls are easily added. RIAA frequency response is within ±0.6 dB of standard values. 0-dB reference gain at 1 kHz is 41.6 dB, producing 1.5-VRMS output from 12.5-mVRMS input. Signal-to-noise ratio is better than −85 dB referenced to 10-mV input level.—"Audio Handbook," National Semiconductor, Santa Clara, CA, 1977, p 2-25–2-31.

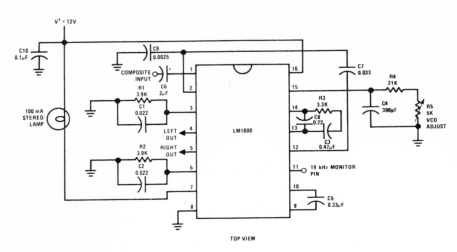

STEREO FM DEMODULATOR—Single National LM1800 IC converts composite AF input signal to left and right signals for audio power amplifiers. LED with series resistor can be used in place of 100-mA lamp.—"Audio Handbook," National Semiconductor, Santa Clara, CA, 1977, p 3-23—3-27.

53-dB PREAMP—RCA CA3052 quad AC amplifier serves for both channels of complete stereo preamp. Circuit is duplicated for other channel. Total harmonic distortion at 1-kHz reference and 1-V output is less than 0.3%. Gain at 1 kHz is 47 dB, with 11.5-dB boost at 100 Hz and 10 kHz. Cut at 100 Hz is 10 dB and at 10 kHz is 9 dB. Operates from single-ended supply. Inputs can be from tape recorders and magnetic-cartridge phonographs.—"Linear Integrated Circuits and MOS/FET's," RCA Solid State Division, Somerville, NJ, 1977, p 327—330.

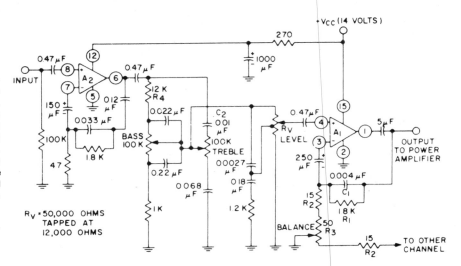

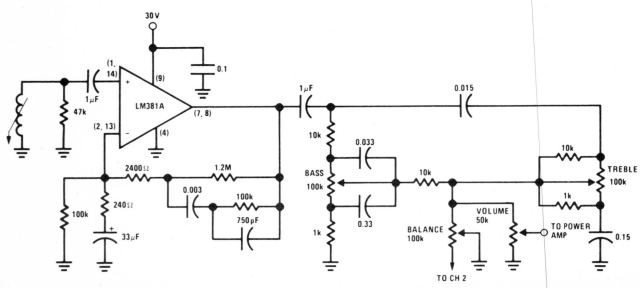

PREAMP WITH TONE CONTROLS—Use of LM381A selected low-noise preamp with passive bass and treble tone controls as phono or tape preamp gives superior noise performance while eliminating need for transistor to offset signal loss in passive controls. Circuit provides 20-dB boost and cut at 50 Hz and 10 kHz relative to midband gain. Design equations are given. Use log pots for tone controls. Other stereo channel is identical. Controls are ganged.—"Audio Handbook," National Semiconductor, Santa Clara, CA, 1977, p 2-40—2-49.

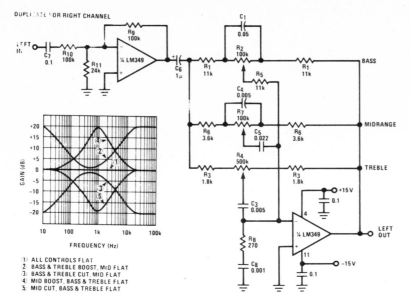

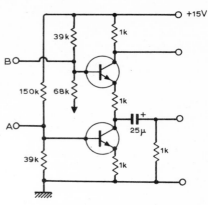

ACTIVE MIDRANGE TONE CONTROL—Addition of midrange tone control to active bass and treble control gives greater control flexibility. Center frequency of midrange control is determined by C_4 and C_5 and is 1 kHz for values shown. C_5 should have 5 times value of C_4.—"Audio Handbook," National Semiconductor, Santa Clara, CA, 1977, p 2-40–2-49.

SUM AND DIFFERENCE—Simple circuit using two BC109 or equivalent transistors is effective for summing and differencing two signals, as required in stereo and quadraphonic sound applications. For resistor values shown, upper output is $-\frac{1}{2}(A + B)$ and lower output is $-\frac{1}{2}(A - B)$. Will handle input signals up to 1.4 V. Bottom of 68K resistor should go to ground.—B. J. Shelley, Active Sum and Difference Circuit, *Wireless World*, July 1974, p 239.

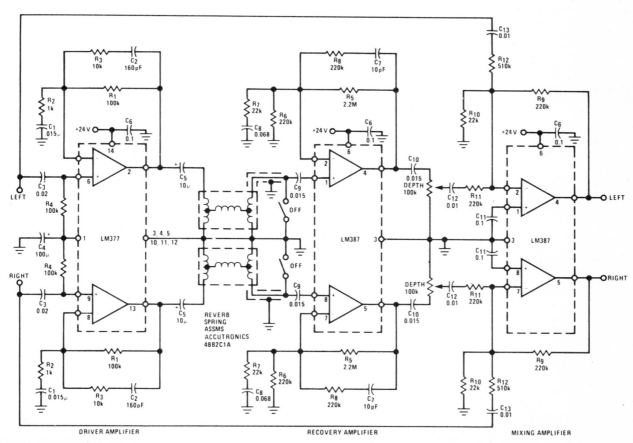

STEREO REVERBERATION—Uses National LM377 dual power amplifier as driver for springs acting as mechanical delay lines. Used to enhance performance of stereo music system by adding artificial reverberation to simulate reflection and re-reflection of sound off walls, ceiling, and floor of listening environment. Amplifier has frequency response of 100–5000 Hz, with rolloff below 100 Hz to suppress booming. Recovery amplifier uses LM387 low-noise dual preamp, and another LM387 provides mixing of delayed signal with original in inverting summing configuration. Output is about half of original signal added to all of delayed signal.—"Audio Handbook," National Semiconductor, Santa Clara, CA, 1977, p 5-7–5-10.

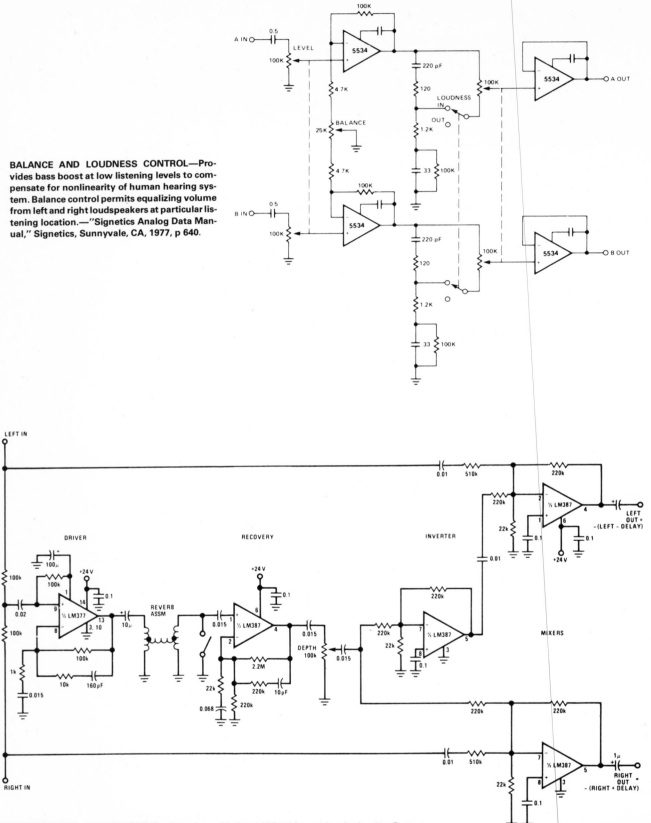

BALANCE AND LOUDNESS CONTROL—Provides bass boost at low listening levels to compensate for nonlinearity of human hearing system. Balance control permits equalizing volume from left and right loudspeakers at particular listening location.—"Signetics Analog Data Manual," Signetics, Sunnyvale, CA, 1977, p 640.

REVERBERATION ENHANCEMENT—Can be used to synthesize stereo effect from monaural source or can be added to existing stereo system. Requires only one spring assembly, which can be Accutronics 4BB2C1A. All opamps are National LM387 low-noise dual units. Outputs are inverted scaled sums of original and delayed signals; left output is left signal minus delay, while right output is right signal plus delay. With mono source, both inputs are tied together and outputs become input minus delay and input plus delay.—"Audio Handbook," National Semiconductor, Santa Clara, CA, 1977, p 5-7—5-10.

ACTIVE TONE CONTROLS USING FEEDBACK—Variation of Baxandall negative-feedback tone control circuit reduces number of capacitors required. Developed for stereo systems. R_4 and R_5 provide negative input bias for opamp, while C_0 prevents DC voltages from being fed back to tone control circuit. For other supply voltages, R_4 is only resistor changed; design procedure is given.—"Audio Handbook," National Semiconductor, Santa Clara, CA, 1977, p 2-40–2-49.

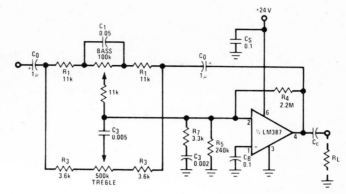

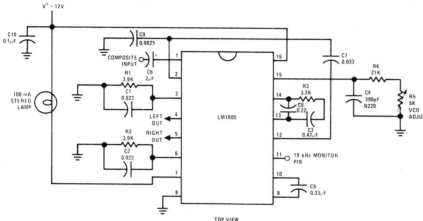

FM DEMODULATOR—National LM1800 PLL IC accepts composite IF output and converts it to separate audio signals for left and right channels. C8 has effect of shunting phase jitter to minimize channel separation problems. If free-running frequency of VCO is set at precisely 19 kHz with R5, separation remains constant over wide range of composite input levels, signal frequencies, temperature changes, and drift in component values.—"Linear Applications, Vol. 2," National Semiconductor, Santa Clara, CA, 1976, AN-81, p 7–8.

LOUDSPEAKER PHASING—Used to determine correct phasing of loudspeakers, microphones, amplifiers, and audio lines in complex stereo systems. Transmitter input feeds sawtooth waveform into stereo input jack of one channel, and receiver unit having microphone input and zero-center meter output is held in front of each loudspeaker in turn for same channel. Components are correctly phased when meter deflects in same direction for all loudspeakers. Procedure is then repeated for other channel. Sawtooth waveform is generated by Analog Devices AD537JD voltage-to-frequency converter. Microphone can be that used with portable cassette recorder. 741 opamp IC$_1$ with gain of 200 feeds dual peak detector D1-D2. Filtered DC signals are detected ramp and detected spike, with spike overriding ramp. Resulting DC level is amplified by 741 opamp having gain of 10, for driving meter. Microphones to be phased are plugged into J1 and connections noted for giving correct meter deflection. J2 is used for phasing amplifiers, lines, and other audio components. Article covers calibration and use.—C. Kitchin, Build an Audio Phase Detector, *Audio*, Jan. 1978, p 54 and 56–57.

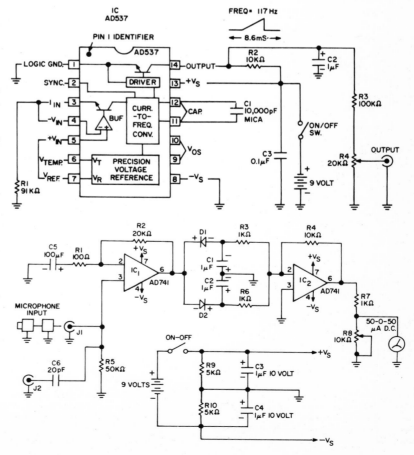

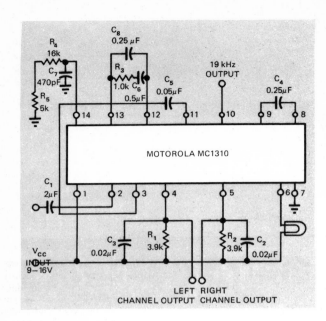

PLL DECODER—Motorola MC1310 phase-locked loop stereo decoder requires only one adjustment, by 5K pot R_5. With pin 2 open, adjust R_5 until reading of 19.00 kHz is obtained with frequency counter at pin 10. Alternatively, tune to stereo broadcast and adjust R_5 to center of lock-in range of stereo pilot lamp. Circuit gives 40-dB separation and about 0.3% total harmonic distortion.—B. Korth, Phase-Locked Loop Stereo Decoder Is Aligned Easily, *EDN Magazine*, Jan. 20, 1973, p 95.

PLL STEREO FM DEMODULATOR—National LM1800 IC uses phase-locked loop techniques to regenerate 38-kHz subcarrier. Automatic stereo/monaural switching is included. Supply voltage range is 10–18 V.—"LM1800 Phase Locked Loop FM Stereo Demodulator," National Semiconductor, Santa Clara, CA, 1974.

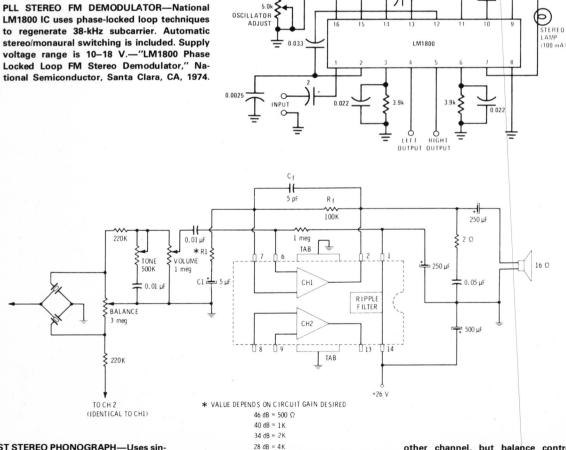

LOW-COST STEREO PHONOGRAPH—Uses single Sprague ULN-2277 IC containing two audio amplifiers each capable of driving loudspeaker directly, for input from high-impedance stereo cartridge. Connections are identical for other channel. Power output per channel is 2 W. Tone and volume controls are ganged with those for other channel, but balance control shown serves both channels.—E. M. Noll, "Linear IC Principles, Experiments, and Projects," Howard W. Sams, Indianapolis, IN, 1974, p 237–239.

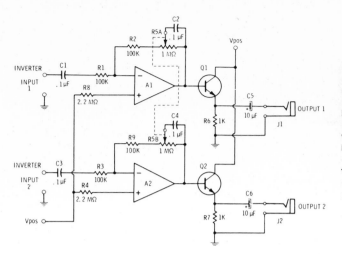

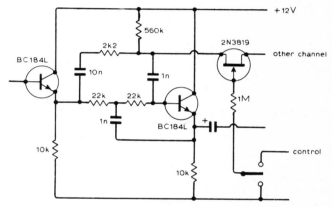

HEADPHONE AMPLIFIER—Designed to drive medium- to high-impedance headphones. Add matching transformers having 1000-ohm primaries if using low-impedance headphones. Dual 1-megohm pot controls gain in stereo channels over range of 1 to 100. Use 9–15 V well-filtered supply rated at least 20 mA. Use Motorola MC3401P or National LM3900 quad opamp and 2N2924 or equivalent NPN transistors.—C. D. Rakes, "Integrated Circuit Projects," Howard W. Sams, Indianapolis, IN, 1975, p 21–24.

FM HISS LIMITER—Uses low-pass filter to remove noise sometimes heard with weak passages during stereo reception of FM stations. FET driven by output of amplifier or tuner is used to switch low-pass filter into operation rather than switching over to mono. Based on fact that the hiss is an antiphase effect that can be removed with little detriment to overall signal.—G. Hibbert, Stereo Noise Limiter Improvement, *Wireless World,* March 1976, p 62.

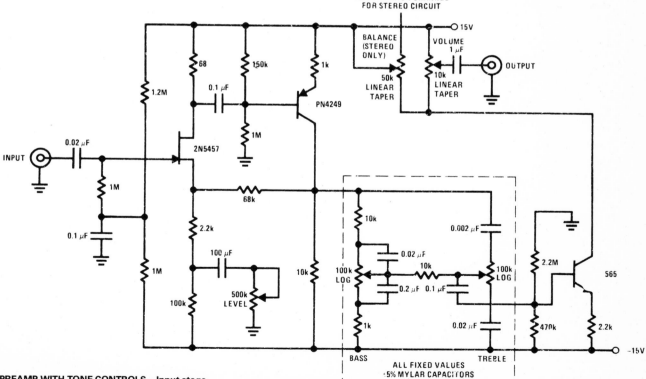

PREAMP WITH TONE CONTROLS—Input stage is JFET having high input impedance and low noise. Circuit parameters are not critical, yet harmonic distortion level is less than 0.05% and S/N ratio is over 85 dB. Tone controls allow 18 dB of cut and boost. Input of 100 mV gives 1-V output at maximum level. Identical preamp is used for other stereo channel.—"FET Databook," National Semiconductor, Santa Clara, CA, 1977, p 6-26–6-36.

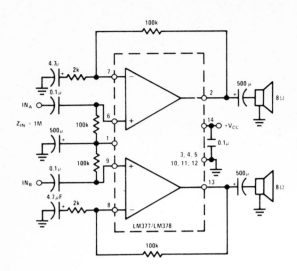

NONINVERTING POWER AMPLIFIER—Single National LM377/LM378 provides gain of 50 and 3 W per channel for driving loudspeakers. Supply is 24 V. High input impedance permits use of high-impedance tone and volume controls. Heatsink is required.—"Audio Handbook," National Semiconductor, Santa Clara, CA, 1977, p 4-8—4-20.

INVERTING POWER AMPLIFIER—Single National LM377 IC provides 2 W per channel with 18-V supply for driving loudspeakers when fed by stereo demodulator of FM receiver. Similar LM378 chip gives 3 W per channel with 24-V supply, and LM379 gives 4 W per channel with 28-V supply. Gain is 50 for all. Heatsink is required.—"Audio Handbook," National Semiconductor, Santa Clara, CA, 1977, p 4-8—4-20.

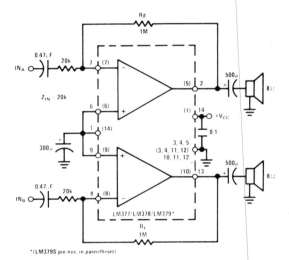

CHAPTER 18
Tape Recorder Circuits

Includes interface circuits for recording and playback of instrumentation and microprocessor data signals, Morse code, and RTTY signals on inexpensive cassette deck, along with NAB-equalized preamps, erase/bias oscillator, AVC, dynamic range expansion, and VOX circuits for all types of mono and stereo tape recorders. Interface for keying CW transmitter with taped message is also given.

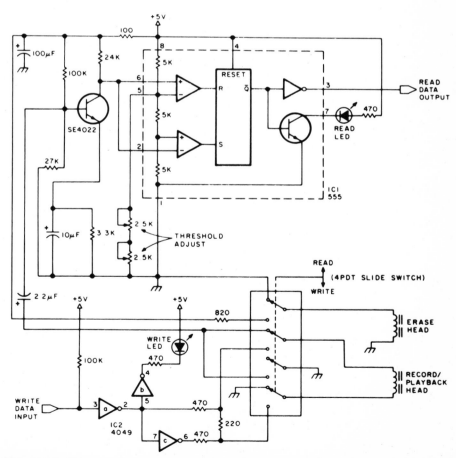

DIGITAL RECORDING WITH CASSETTES—Circuit shows modifications required for standard cassette recorder to bring read level up to about 1 V. Recorder works well over range of 100 to 1200 b/s. During write process, direct current is passed through record head to saturate tape, with polarity depending on direction of current. During read cycle, voltage is induced in head winding only when transition between oppositely polarized zones moves past head. 555 timer is used as combination level detector and flip-flop to recover serial data.—R. W. Burhans, A Simpler Digital Cassette Tape Interface, *BYTE,* Oct. 1978, p 142–143.

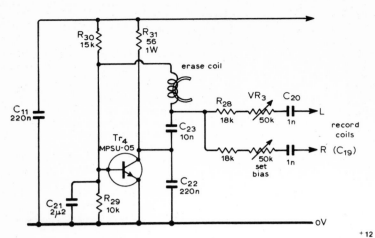

ERASE/BIAS OSCILLATOR—Used in high-quality stereo cassette deck operating from AC line or battery. Provides up to 33 VRMS at 50-kHz erase frequency, as required for completely erasing existing recording on tape when recording over it. Supply voltage should be in range of 12–14 V. Article gives all other circuits of cassette deck and describes operation in detail.— J. L. Linsley Hood, Low-Noise, Low-Cost Cassette Deck, *Wireless World*, Part 1—May 1976, p 36–40 (Part 2—June 1976, p 62–66; Part 3—Aug. 1976, p 55–56).

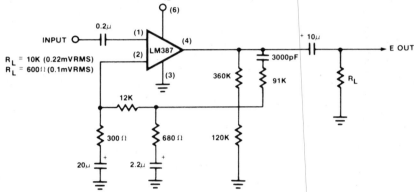

CASSETTE PREAMP—Provides gain of 81 dB and 0.22 mVRMS for 10K load. Gain drops to about 78 dB and output is 0.1 mVRMS for 600-ohm load. Gain values are for 100 Hz, with gain dropping above and below this value.—"Signetics Analog Data Manual," Signetics, Sunnyvale, CA, 1977, p 782.

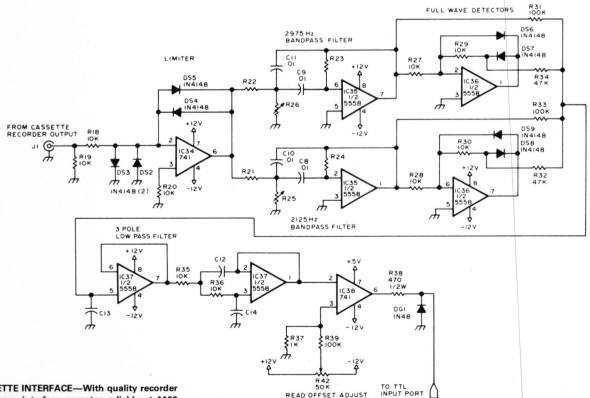

CASSETTE INTERFACE—With quality recorder and tapes, interface operates reliably at 1100 bauds, for loading 24K microprocessor system in 246 s. Cassette output is amplified and clipped by limiting amplifier IC34. Bandpass filters followed by full-wave detectors respond to 2125-Hz mark and 2975-Hz space frequencies and feed their outputs to summing junction at pin 5 of three-pole active low-pass filter IC37. 2975-Hz tones are rectified to positive voltage and 2125-Hz tones to negative voltage, with amplitudes varying from maximum at exact frequencies to sum voltage of 0 V at midfrequency of 2550 Hz. Output opamp IC38 delivers correct TTL level for reading by single-bit input port.— R. Suding, Why Wait? Build a Fast Cassette Interface, *BYTE*, July 1976, p 46–53.

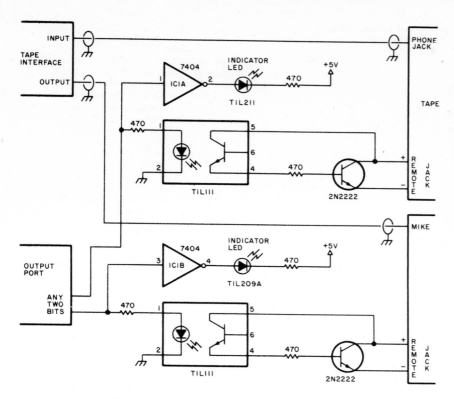

CASSETTE FILE UPDATE—Interface circuit controls two tape decks for updating mailing lists or other sequential files stored on magnetic tape in audio cassettes. Two cassette tape recorders are required, one for input (reading files) and one for output (writing files). Microprocessor tape input and output circuits are connected to appropriate tape unit as shown. Only one cassette operates at any given time. Optocouplers prevent polarity or voltage problems between tape motor and microprocessor. Tape functions are under software control. Software delay of about 1 s allows tape motor to come up to speed before recording starts. Records are in numerical or other logical sequence, so updating requires only one pass. On update, old cassette file is read into microprocessor for deletion, change, or addition of data, and corrected data is written on new cassette. Article covers use for maintaining Christmas card and other mailing lists, payroll records sequenced by Social Security number, and other sequential lists.—W. D. Smith, Fundamentals of Sequential File Processing, *BYTE,* Oct. 1977, p 114–116, 118, 120, 122, 124, and 126–127.

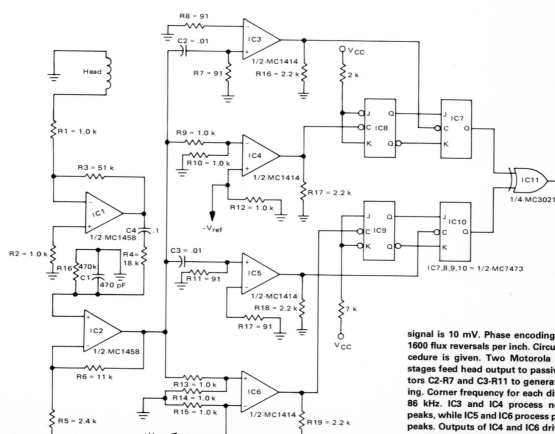

CASSETTE DATA READOUT—Uses separate circuits having threshold provisions for both positive and negative peaks, for reading data stored on cassette tape at 15 in/s. Head output signal is 10 mV. Phase encoding is used with 1600 flux reversals per inch. Circuit design procedure is given. Two Motorola MC1458 gain stages feed head output to passive differentiators C2-R7 and C3-R11 to generate zero crossing. Corner frequency for each differentiator is 86 kHz. IC3 and IC4 process negative-going peaks, while IC5 and IC6 process positive-going peaks. Outputs of IC4 and IC6 drive T flip-flops serving as data inputs to IC7 and IC10.—"The Recovery of Recorded Digital Information in Drum, Disk and Tape Systems," Motorola, Phoenix, AZ, 1974, AN-711, p 9.

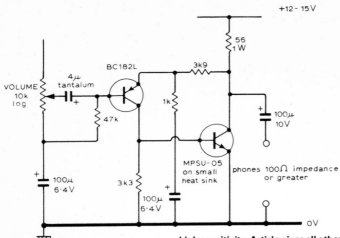

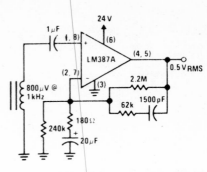

NAB PLAYBACK PREAMP—Provides standard NAB equalization for tape player requiring 0.5 VRMS from head having sensitivity of 800 μV at 1 kHz, with operating speed of 3¾ in/s. Design procedure is given. Voltage gain at 1 kHz is 56 dB.—"Audio Handbook," National Semiconductor, Santa Clara, CA, 1977, p 2-31–2-37.

HEADPHONE AMPLIFIER—Used in high-quality stereo cassette deck operating from AC line or battery. Provides gain of 5 in class A, for use with low-sensitivity headphones or low-impedance headphones down to 100 ohms. Replay amplifier output alone is adequate for headphones having 2000-ohm load impedance or high sensitivity. Article gives all other circuits of cassette deck and describes operation in detail. Input to volume control is taken from output of opamp in replay amplifier, nominally about +5 V.—J. L. Linsley Hood, Low-Noise, Low-Cost Cassette Deck, *Wireless World*, Part 2—June 1976, p 62–66 (Part 1—May 1976, p 36–40; Part 3—Aug. 1976, p 55–56).

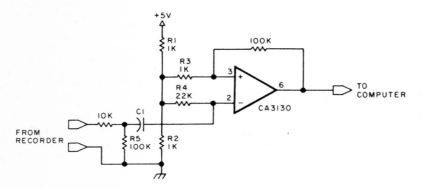

CASSETTE INTERFACE—Used between recorder and computer for loading data stored in tape cassette. Single divider network R1-R2 drives both opamp inputs and provides stabilized sensitivity. R3 isolates inputs.—B. E. Rehm, The TDL System Monitor Board, *BYTE*, April 1978, p 10, 12–14, and 16.

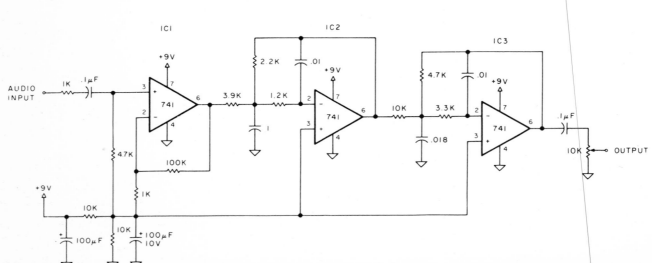

COPYING CASSETTE PROGRAMS—Controller serves for making duplicate copies of microprocessor programs recorded on magnetic tape, for insurance against accidental damage to master cassette during use. Used between audio output of cassette player and audio input of tape recorder. Opamp IC1 with gain of 100 overloads so output is constant-amplitude square wave regardless of input level from tape being copied. If program uses audio tones for digital data, eight cycles of 2400 Hz represents digital 1 and four cycles of 1200 Hz represents digital 0. Additional opamps act as four-pole Butterworth filter rejecting signals above 3000 Hz. 10K pot is adjusted so output level matches requirements of recorder.—P. A. Stark, Copying Computer Cassettes, *Kilobaud*, Aug. 1978, p 94–96.

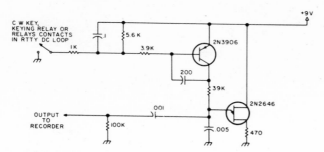

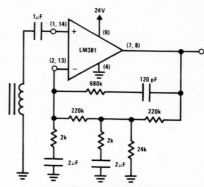

CW AND RTTY ON CASSETTES—Circuit provides conditioning of routine CW calls or RTTY test messages, as required for recording on endless-loop cassette. Keyed signal is filtered to remove contact bounce, then used to turn on

2N3906 which gates 2N2646 sawtooth oscillator operating at about 5 kHz when using 0.005-µF gate capacitor; for lower frequency, increase capacitor to 0.01 µF.—Cassette-Aided CW and RTTY, *73 Magazine*, Sept. 1977, p 122–123.

FAST TURN-ON PLAYBACK PREAMP—Turn-on for gain and supply voltage is only 0.1 s, as compared to 5 s normally required in preamp providing NAB tape playback response.—"Audio Handbook," National Semiconductor, Santa Clara, CA, 1977, p 2-31–2-37.

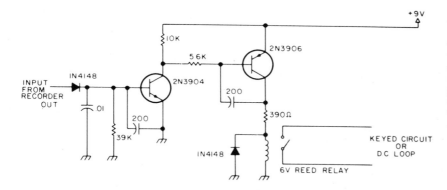

CASSETTE PLAYBACK OF CW AND RTTY—Playback-signal conditioning circuit is used between tape recorder and transmitter when routine CW calls or RTTY test messages are recorded on endless-loop cassette recorder. Recorded tone is rectified by 1N4148 and applied to RC timing circuit. Decay voltage developed across network when tone is removed turns on 2N3904 and 2N3906 stages. Output of 2N3906 drives reed relay in transmitter keying circuit. If resistor is used in place of relay, drop across it during key-down period can be used to drive electronic keyer.—Cassette-Aided CW and RTTY, *73 Magazine*, Sept. 1977, p 122–123.

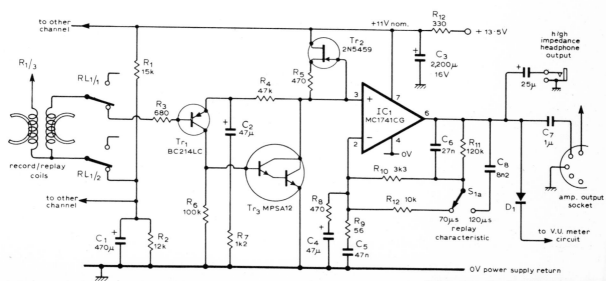

REPLAY AMPLIFIER—Used in high-quality stereo cassette deck operating from AC line or battery. Amplifier design is optimized for minimum noise voltage by using PNP silicon input transistor operated with lowest possible collec-

tor current (10 µA for Texas Instruments transistor specified). Motorola IC in second stage, similar to 741 but having 8-pin metal-can encapsulation, provides equalization required for replay. Output of amplifier is about 0.4 VRMS.

Article gives all other circuits of cassette deck and describes operation in detail.—J. L. Linsley Hood, Low-Noise, Low-Cost Cassette Deck, *Wireless World*, Part 1—May 1976, p 36–40 (Part 2—June 1976, p 62–66; Part 3—Aug. 1976, p 55–56).

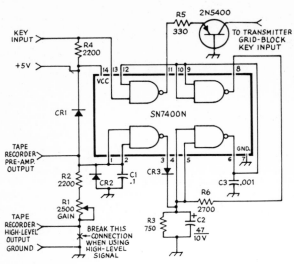

KEYING FROM TAPE—Simple envelope detector and wave-shaping circuit uses quad NAND gate for instant replay of recorded CW transmissions through transmitter. Diodes can be 1N270 or any other small-signal switching or general-purpose types. R3, C2, and CR3 provide envelope detection of amplified and clipped audio input from tape recorder.—A. H. Kilpatrick, Keying a Transmitter with a Tape Recorder, *QST,* Jan. 1974, p 45.

DIGITAL CASSETTE HEAD DRIVE—Provides saturation recording as required for digital data. Back-to-back zeners provide bipolar limiting at ±10 V. TTL-level inputs are applied to write data input, inverted by 7404, and fed to inverting input of opamp. Noninverting opamp input is referenced to +1.4 V so output will switch polarities when TTL level of input changes.—I. Rampil and J. Breimeir, The Digital Cassette Subsystem: Digital Recording Background and Head Interface Electronics, *BYTE,* Feb. 1977, p 24–31.

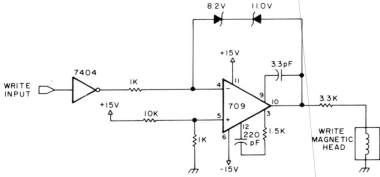

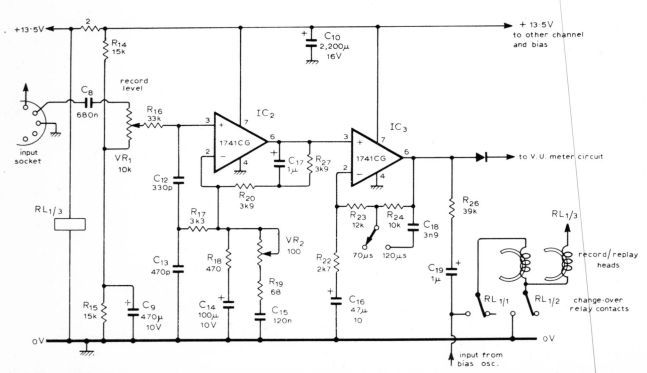

RECORDING AMPLIFIER—Used in high-quality stereo cassette deck operating from AC line or battery. Uses active RC circuit R16-R17-C12-C13-R19-VR2-C15 to provide required high-frequency recording characteristic for use with Garrard CT4 recording head; component values may have to be changed for other heads. C18 (3.9 nF) is switched in to change from basic 70-μs recording characteristic to 120 μs. C17 and R27 provide new cassette-standard bass preemphasis at 3,180 μs. Recording level is chosen as 0 VU at 660 Hz. Output feeds VU meter through silicon diode. Article gives all other circuits of cassette deck and describes operation in detail.—J. L. Linsley Hood, Low-Noise, Low-Cost Cassette Deck, *Wireless World,* Part 1—May 1976, p 36–40 (Part 2—June 1976, p 62–66; Part 3—Aug. 1976, p 55–56).

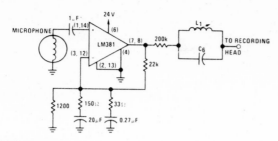

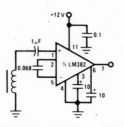

RECORDING AMPLIFIER—Designed for use with microphone having 10-mV peak output and recording head requiring 30-μA AC drive current. Output swing is 6 VRMS. High-frequency cutoff is 16 kHz, with circuit designed for slope of 6 dB per octave between 4 kHz and 16 kHz to compensate for falling frequency response of recording head starting at 4 kHz.—"Audio Handbook," National Semiconductor, Santa Clara, CA, 1977, p 2-31–2-37.

PLAYBACK PREAMP—Circuit is optimized for automotive use at supply of 10–15 V. Wideband 0-dB reference gain is 46 dB. NAB equalization is included. Tape speeds can be 1⅞ or 3¾ in/s.—"Audio Handbook," National Semiconductor, Santa Clara, CA, 1977, p 2-31–2-37.

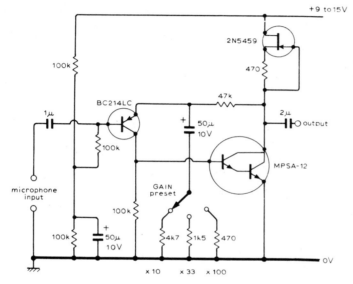

MICROPHONE PREAMP—Used in high-quality stereo cassette deck operating from AC line or battery. Provides three preset gain positions (10, 33, and 100) to meet amplification requirements of practically all types of microphones used with tape recorders. Recording input of cassette deck provides only enough gain for recording from audio amplifier or radio tuner delivering 50–100 mV at fairly low impedance, hence is not suitable for microphone input. Article gives all other circuits of cassette deck and describes operation in detail.—J. L. Linsley Hood, Low-Noise, Low-Cost Cassette Deck, *Wireless World*, Part 2—June 1976, p 62–66 (Part 1—May 1976, p 36–40; Part 3—Aug. 1976, p 55–56).

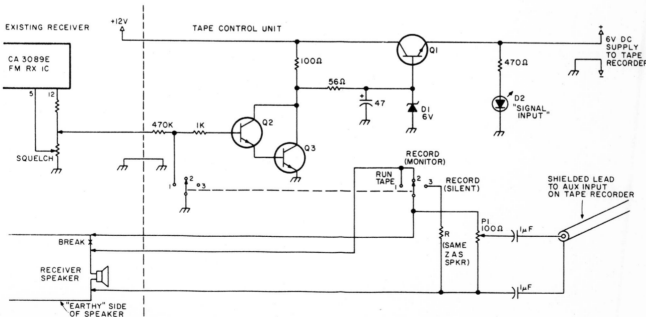

MESSAGE-CONTROLLED RECORDER—Circuit turns on tape recorder whenever input signal is present in receiver, and turns off recorder when signal goes off. Applications include monitoring local FM repeater for daily usage to obtain call signs of users, or unattended recording of messages left by other amateur stations. Uses cheap cassette tape recorder with autostop, operating at 6 V obtained from 12-V receiver supply by series regulator Q1 and zener D1. Connection to mute or squelch circuit of receiver is shown for set having CA3089E in IF tail end. Darlington pair Q2-Q3 effectively removes base supply for Q1 to turn recorder off. LED comes on when recorder is on. Q1 is NPN power transistor, while Q2 and Q3 are small-signal NPN transistors.—F. Johnson, Automatic Taping Unit, *73 Magazine,* May 1977, p 98–99.

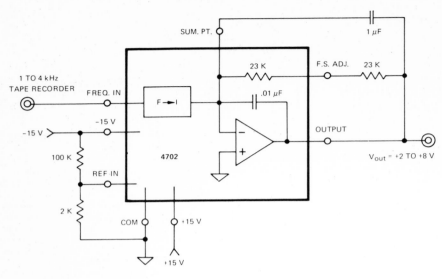

PLAYBACK OF PULSE TRAINS—Teledyne Philbrick 4702 frequency-to-voltage converter circuit provides ripple filter required for converting recorded square waves in frequency range of 0.5 to 5 kHz to desired analog output in range of 2 to 8 VDC. Report covers problems of recording and playing back pulse trains.—"V-F's, F-V's, and Audio Tape Recorders," Teledyne Philbrick, Dedham, MA, 1974, AN-11.

STEREO TAPE PLAYBACK—Single Sprague ULN-2126A IC provides preamplification for two channels along with 2-W output power for driving stereo power amplifier. Values shown give equalization required for tape playback. Single ganged tone control serves for both treble and bass adjustment.—E. M. Noll, "Linear IC Principles, Experiments, and Projects," Howard W. Sams, Indianapolis, IN, 1974, p 237 and 243.

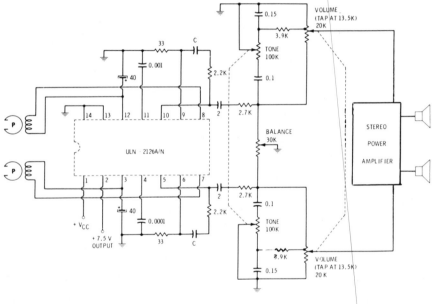

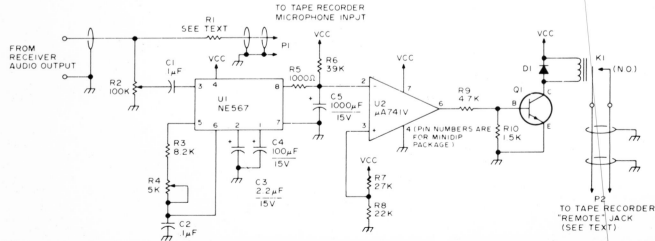

UNATTENDED RECORDER—Uses 567 tone decoder in circuit designed to respond to 1-kHz tone, to turn on recorder for taking message when receiver of amateur station is unattended. R6, C5, and 741 opamp U2 form timer that turns on RS-267-2016 transistor and pulls in relay to turn on tape recorder for recording about 30-s message. Relay then drops out. Use well-regulated 5-V supply. All transmitters using this service must have 1-kHz audio encoders for producing required control frequency. Article gives construction and adjustment details.—R. Perlman, The F.M. "Auto-Start," *73 Magazine,* April 1974, p 21 and 23–24.

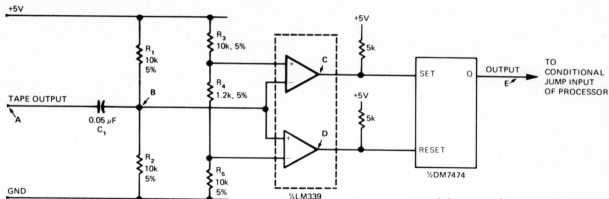

INTERFACE FOR AUDIO CASSETTES—Permits use of ordinary home cassette recorder to provide high-speed loading of assembler and source program into microprocessor. Data is recorded by using variation of phase encoding, which provides self-clocking and is independent of tape speed variation. Effective I/O rate is about 500 b/s or 5 times that of low-speed paper-tape punch or reader. Article covers phase-encoding procedure, gives flowchart, and shows waveforms of pulses at five points in circuit. Parity bits provide error correction and detection, using Hamming code.—S. Kim, An Inexpensive Audio Cassette Recorder Interface for μP's, *EDN Magazine,* March 5, 1976, p 83–86.

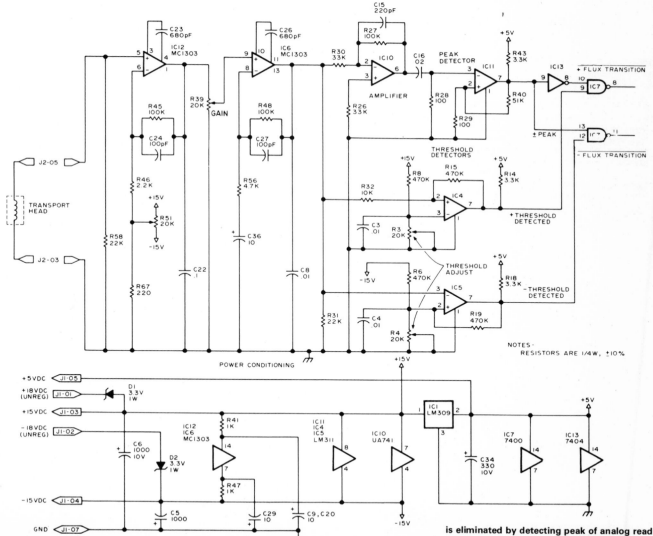

DIGITAL CASSETTE READ AMPLIFIER—Signal from magnetic head of digital tape cassette is amplified by two-stage MC1303 amplifier providing analog output of about 4 V P-P to μA741 opamp IC10 of LM311 peak detector IC11. Signal also goes to LM311 positive and negative threshold detectors IC4 and IC5, which give logic-level output. When input signal is below preset reference level, output of positive threshold detector is low; above reference, output is high. Negative threshold detector operates similarly for negative pulses. Time jitter in outputs is eliminated by detecting peak of analog read signal, then combining result with threshold information in peak detector. Circuit is used in Phi-Deck cassette system made by Economy Company, Oklahoma City.—I. Rampil and J. Breimeir, The Digital Cassette Subsystem: Digital Recording Background and Head Interface Electronics, *BYTE,* Feb. 1977, p 24–31.

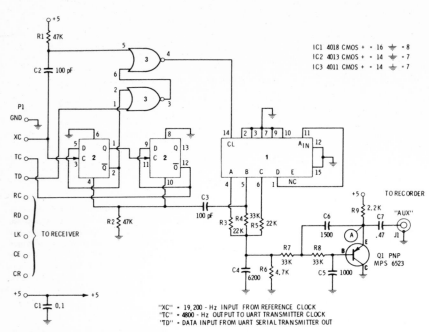

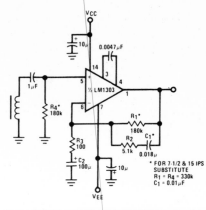

IC1 4018 CMOS + = 16 = 8
IC2 4013 CMOS + = 14 = 7
IC3 4011 CMOS + = 14 = 7

"XC" = 19,200 - Hz INPUT FROM REFERENCE CLOCK
"TC" = 4800 - Hz OUTPUT TO UART TRANSMITTER CLOCK
"TD" = DATA INPUT FROM UART SERIAL TRANSMITTER OUT

FOUR-SPEED PLAYBACK PREAMP—Provides 0-dB reference gain of 34 dB. Supply can be ±4.5 to ±15 V. Values shown are for NAB equalization and 1⅞ or 3¾ in/s; for 7½ and 15 in/s, change values as indicated. Design equations are given.—"Audio Handbook," National Semiconductor, Santa Clara, CA, 1977, p 2-31–2-37.

300-BAUD BIT BOFFER TRANSMITTER—Permits recording of serial data on ordinary low-cost cassette tape recorders for bulk storage of data to be used later in microprocessor. Requires 19,200-Hz reference input to terminal XC from external clock. Feedback from sine-wave synthesizer IC1 to divide-by-4 counter IC2 automatically synchronizes system so sine waves automatically switch just before zero crossing each time serial data changes from 1 to 0 or back again. Output consists of 16 half sine waves at 2400 Hz for mark or digital 1 and 8 half sine waves at 1200 Hz for space or digital 0, for feed to input of cassette recorder. Circuit eliminates errors commonly encountered when attempting to record square waves on tape with low-cost recorder.—D. Lancaster, "TV Typewriter Cookbook," Howard W. Sams, Indianapolis, IN, 1976, p 167–171.

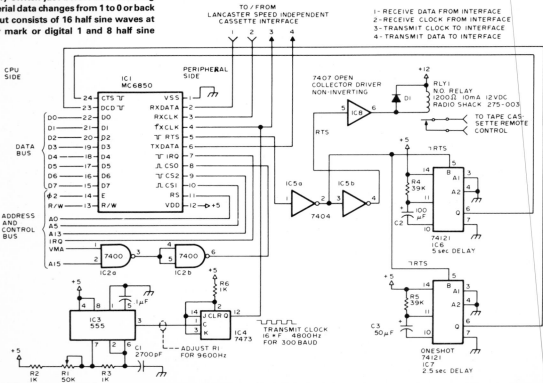

TO / FROM
LANCASTER SPEED INDEPENDENT
CASSETTE INTERFACE

1- RECEIVE DATA FROM INTERFACE
2- RECEIVE CLOCK FROM INTERFACE
3- TRANSMIT CLOCK TO INTERFACE
4- TRANSMIT DATA TO INTERFACE

CASSETTE INTERFACE WITH ACIA—Permits use of audio pickup for mass storage in Motorola 6800 microcomputer system. Uses Motorola MC6850 asynchronous communication interface adapter (ACIA), which is specialized version of UART. All control, status, and data transfers in ACIA are made over single 8-bit bidirectional bus. Request-to-send line (RTS) controls tape recorder motor. When RTS is set high, input to IC8 is high and relay coil is not energized. IC6 gives 5-s delay following motor turn-on so long leader will be recorded at mark frequency. IC7 gives delay so reading starts 2.5 s before first data byte. Article covers circuit operation in detail and gives operating subroutines.—J. Hemenway, The Compleat Tape Cassette Interface, *BYTE*, March 1976, p 10–16.

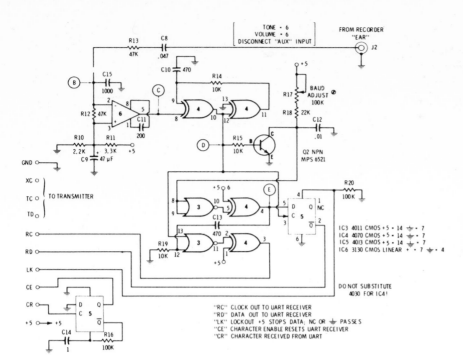

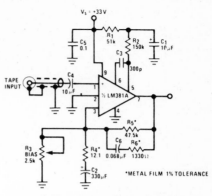

ULTRALOW-NOISE PLAYBACK PREAMP—Provides optimum noise performance at popular tape speeds of 1⅞ and 3¾ in/s. Reference gain for 0 dB is 41 dB, giving output level of 200 mV from head output of 1 mV at 1 kHz. Single-ended biasing and use of metal-film resistors reduce noise.—"Audio Handbook," National Semiconductor, Santa Clara, CA, 1977, p 2-31–2-37.

300-BAUD BIT BOFFER RECEIVER—Used with ordinary cassette recorder to convert half sine waves of recorded serial data to corresponding digital 1s and 0s. Output of recorder passes through filter and limiter IC6 to give square wave at point C whose zero crossings correspond to recorded sine wave. Leading and trailing edges of square wave are converted to narrow positive pulses by EXCLUSIVE-OR gate IC4 to give stream of pulses at D, one for each zero crossing. Transistor circuit forms retriggerable mono that is adjusted so point E goes positive three-fourths of way through low-frequency (1200 Hz) half-cycle. Point E then has stream of eight pulses for 0 and no pulses for 1. Final flip-flop provides recovery of data as 1s and 0s. Leading edge of waveform at D is shortened and combined with clock pulses to provide composite UART clock output. Boffer system eliminates errors commonly encountered when attempting to record square waves on tape with low-cost recorder.—D. Lancaster, "TV Typewriter Cookbook," Howard W. Sams, Indianapolis, IN, 1976, p 167–171.

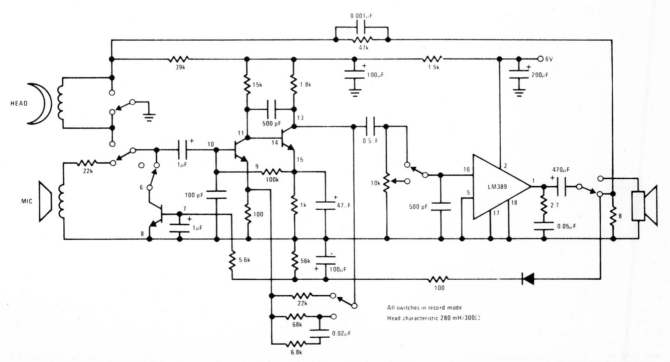

CASSETTE RECORD/PLAYBACK—National LM389 power amplifier chip includes three NPN transistors, to provide all circuits needed for complete recording and playback of cassette tapes. Two of internal transistors act as signal amplifiers while third is used for automatic level control when recording. Diode is also on chip.—"Audio Handbook," National Semiconductor, Santa Clara, CA, 1977, p 4-33–4-37.

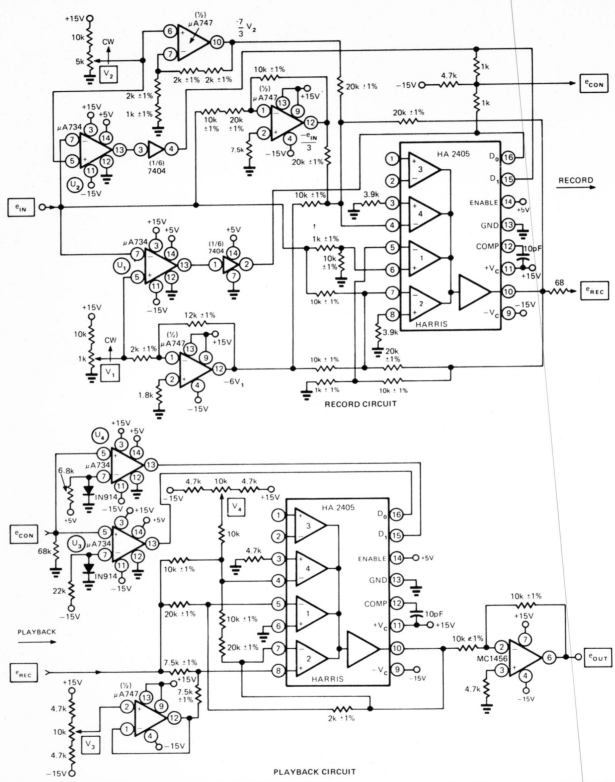

RECORD CIRCUIT

PLAYBACK CIRCUIT

AUTOMATIC RANGE EXPANSION—Instrumentation tape recorder technique folds recorded signal over and reuses same VCO range three times, at three different gains, for increasing dynamic recording range to over 10,000. Two comparators select one of three amplifier gains according to level of input signal and record selected gain on separate control track. During playback, control track signal e_{CON} is used to select corresponding inverse gain for unfolding recorded signal. Level of input signal e_{IN}, in range of 0–10 V, is sensed by comparators whose preset thresholds are determined by pots V_1 and V_2. If input is less than V_1, both comparator outputs are low and section 1 of HA2405 four-channel opamp is selected for recording at 10 times input. If input is greater than V_1 and less than V_2, section 2 having gain of −2 is selected so direction of e_{REC} is reversed. If e_{IN} is greater than V_2, both comparators are high and section 4 is selected for gain of +1/3, so e_{REC} again reverses to cross VCO range for third time. Outputs of comparators are summed to form three-level signal for recording on control track.—J. R. White, Comparator Technique Expands Tape Recorder's Range, *EDN Magazine*, April 5, 1975, p 111, 113, and 115.

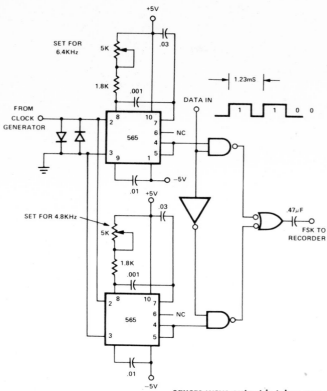

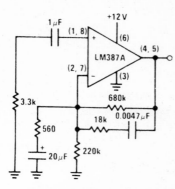

12-V PLAYBACK PREAMP—Provides standard NAB equalization. Gain is decreased gradually from 60 dB at 20 Hz to 32 dB at 20 kHz in accordance with NAB playback curve. Playback head is represented by 3.3K resistor.—"Audio Handbook," National Semiconductor, Santa Clara, CA, 1977, p 2-31–2-37.

FSK GENERATOR FOR CASSETTE DATA—Uses two 565 PLL ICs, locked to 800-Hz system clock but oscillating at 6.4 kHz and 4.8 kHz, to provide FSK signals for recording digital data on ordinary cassette tape. Harmonic suppression of square-wave output is taken care of automatically by high-frequency rolloff characteristic of tape recorder. Incoming data determines which oscillator feeds its signal to recorder.—"Signetics Analog Data Manual," Signetics, Sunnyvale, CA, 1977, p 859–860.

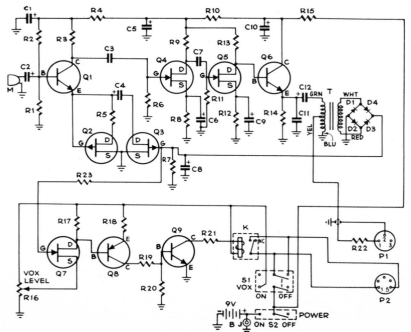

C1, 4 8, 10, 12—50mfd electrolytic capacitor
C2—2.2mfd electrolytic capacitor
C3, 7, 11—0.1mfd ceramic capacitor
C5—220mfd electrolytic capacitor
C6, 9—100mfd electrolytic capacitor
R1—120,000-ohm, ¼w resistor
R2—56,000-ohm, ¼w resistor
R3, 9, 13—10,000-ohm, ¼w resistor
R4, 10, 14—1000-ohm, ¼w resistor
R5—15,000-ohm, ¼w resistor
R6, 11—1 megohm, ¼w resistor
R7—120,000-ohm, ¼w resistor
R8, 12—2700-ohm, ¼w resistor
R15—56-ohm, ¼w resistor
R16—50,000-ohm, miniature potentiometer
R17—6800-ohm, ¼w resistor
R18—1500-ohm, ¼w resistor
R19—150-ohm, ¼w resistor
R20—2200-ohm, ¼w resistor
R21—330-ohm, ¼w resistor
R22, 23—560,000-ohm, ¼w resistor
D1, 2, 3, 4—diodes 1N266 or equiv.
Q1-Q6—NPN transistor Motorola HEP 50
Q2, 3—P-channel FET 2N3820
Q4, 5, 7—N-channel FET Motorola HEP 801
Q8—PNP transistor Motorola HEP 52
Q9—NPN transistor Motorola HEP 53
T—output transformer 1K-200K Radio Shack
    273-1376
K—miniature relay Radio Shack 275-004

AVC AND VOX—Voice-operated ON/OFF switch uses microphone to sense normal background sound. Anything above background threshold preset by R16 energizes relay K for turning on recorder. Circuit provides about 2-s delay after subject stops talking, before releasing relay. Automatic volume control circuit keeps recorded signal essentially constant despite movements of loudspeaker toward or away from microphone.—G. Beard, Automatic Volume and VOX for Your Tape Recorder, *Popular Science*, Oct. 1973, p 134 and 136.

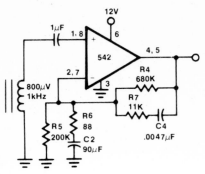

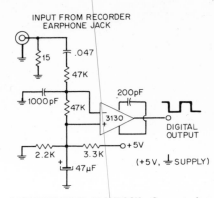

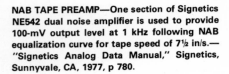

NAB TAPE PREAMP—One section of Signetics NE542 dual noise amplifier is used to provide 100-mV output level at 1 kHz following NAB equalization curve for tape speed of 7½ in/s.—"Signetics Analog Data Manual," Signetics, Sunnyvale, CA, 1977, p 780.

CASSETTE DATA PLAYBACK—Converts low-level digital signals from cassette recorder into CMOS-compatible 5-V square waves. Both inputs of 3130 opamp are biased to +2 V for use as open-loop comparator. RC input filter minimizes hum and bias interference.—D. Lancaster, "CMOS Cookbook," Howard W. Sams, Indianapolis, IN, 1977, p 345.

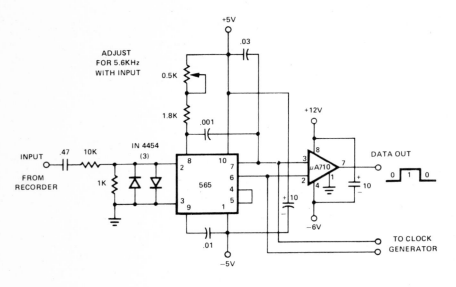

RECORDED FSK

OUTPUT ON PIN 7

FSK DETECTOR FOR CASSETTE-RECORDED DATA—Connection shown for 565 PLL provides data output of 1 for 6.4 kHz and 0 for 4.8 kHz from ordinary cassette tape recorder having frequency response to 7 kHz. Report gives circuit of suitable recorder using return-to-zero FSK. System also requires 800-Hz clock generator for synchronizing to data. Up to seven 0s can occur in succession without making clock go out of sync. Odd parity is used.—"Signetics Analog Data Manual," Signetics, Sunnyvale, CA, 1977, p 857–859.

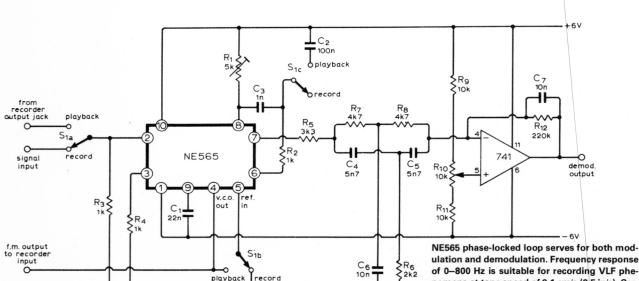

FM FOR INSTRUMENTATION—Frequency modulator-demodulator circuit using single IC and opamp converts ordinary low-cost tape recorder into instrumentation recorder. Signetics NE565 phase-locked loop serves for both modulation and demodulation. Frequency response of 0–800 Hz is suitable for recording VLF phenomena at tape speed of 9.1 cm/s (3.5 in/s). Carrier frequency is in midband, at 3 kHz. Article covers circuit operation in detail and gives design equations.—B. D. Jordan. Simple F.M. Modulator/Demodulator for a Magnetic Tape Recorder, *Wireless World*, March 1974, p 29–30.

Name Index

NOTE: The persons named in this index are, in most instances, authors of articles cited in this book.

Subject Index